Enjoy the book!

Most of the diagrams can be downloaded from

www.mathematicstodo.co.uk.

Chris O'Donoghue

MATHEMATICS TO DO

MATHEMATICS TO DO

A Recreational Mathematics Book

Chris O'Donoghue

The Book Guild Ltd
Sussex, England

The Book Guild Ltd.
25 High Street,
Lewes, Sussex

First published 2000

Set in Times
Typesetting by
Acorn Bookwork, Salisbury, Wiltshire

Printed in Great Britain by
Bookcraft (Bath) Ltd, Avon

A catalogue record for this book is
available from the British Library

ISBN 1 85776 469 2

CONTENTS

PREFACE

There are many excellent books devoted to recreational mathematics, and if you are a good mathematician, they are fun to read. But what if you are not such a good mathematician, but still have an interest in the subject?

A short time ago, I offered to tutor a recreational mathematics course at my local University of the Third Age. My students had no particular mathematical expertise but plenty of enthusiasm and sufficient interest in the subject to spend an hour and a half a week studying the course, and goodness only knows how much time at home completing each chapter.

My challenge was to find material suitable for these enthusiasts. I could find none in any existing books, so the only solution was to write it, and here it is!

I am grateful to my students for their encouragement, because unlike a school or college course, they can leave at any time if they consider the work to be too difficult or not sufficiently interesting. They have politely corrected my errors, and let me know by the bewildered expressions on their faces when I have not explained something very well.

My plan has been to investigate branches of mathematics which I consider to be interesting either intrinsically or because of their application to something in the 'real' world. From my many years of school teaching, I know that the only way anyone learns mathematics (or anything else?) is to do it! Consequently, instead of paragraphs of explanation which you would find in most recreational mathematics books, you are presented here with a series of questions which, hopefully, help you to see the mathematical point of the exercise. The questions are *not* for routine practice or to fix an idea firmly in your mind as might be the case in a normal textbook; they are there by way of explanation.

I have tried to cater for a wide ability range in the questions, and although most of my students were not mathematicians, many of them were determined to master everything which I presented to them. However, I separated some of the more difficult work into an 'Extra' section at the end of certain chapters so that the less confident would not mind leaving some work undone. I have always thought that it is algebra which intimidates most people when facing up to mathematics, and it is mostly algebra, as well as some more difficult ideas, which I have confined to the 'Extra' section.

Although the material in this book was written originally for over-50-year-olds, and some were over 70, there is no lower age limit to its readers. I would think that under-13s would have to be pretty bright to manage all of it. Much of the content of this book is off the current schools' syllabus, but where syllabus topics have been included, I have tried to give an unusual slant on the work.

For those of you who are 'real' mathematicians, you may find some of the work here does not have the mathematical rigour you would like. Remember that this is not written for 'real' mathematicians, but for those of us who enjoy the subject in our own way.

1

The Golden Section

For this chapter you will need:

A ruler which can measure in millimetres, pen and paper and a calculator
A photocopy or tracing of figures 1 and 4
A compass (correctly called a pair of compasses!)

This chapter introduces the Golden Section as the ratio of the length of a side of a regular pentagon to the length of one of its diagonals. As you work your way through the chapter, I hope you will begin to realise what the Golden Section has to do with our lives.

We see the Golden Section introduced the way the Greeks probably knew it. They would not have been able to recognise it as 1:1.618... because they did not know decimals, and the number 1.618... is irrational; that is it cannot be expressed accurately as a fraction (or as a decimal come to think of it!). However, the mathematicians of Ancient Greece worked with ratios, and this was one which fascinated them.

The faint-hearted – a polite term for those who are not very good mathematicians – should not go on to the Extra section; they never should!

The first part of the Extra section is there to lead into Fibonacci Numbers, which are dealt with in the second chapter. If you think of it, look back at this work when you have finished Chapter 2.

The next part of the Extra section shows how the value 1.618... is calculated and requires confidence and competence in algebra sufficient to solve a quadratic equation which will not factorise.

The demonstration of why $G^2 = G + 1$ and $\frac{1}{G} = G - 1$ requires a good ability in algebraic manipulation, but I have to put in something for the good mathematicians.

The final part of this section is easy enough if you are good at geometry, but a great deal – as textbook writers always say – has been left to the student.

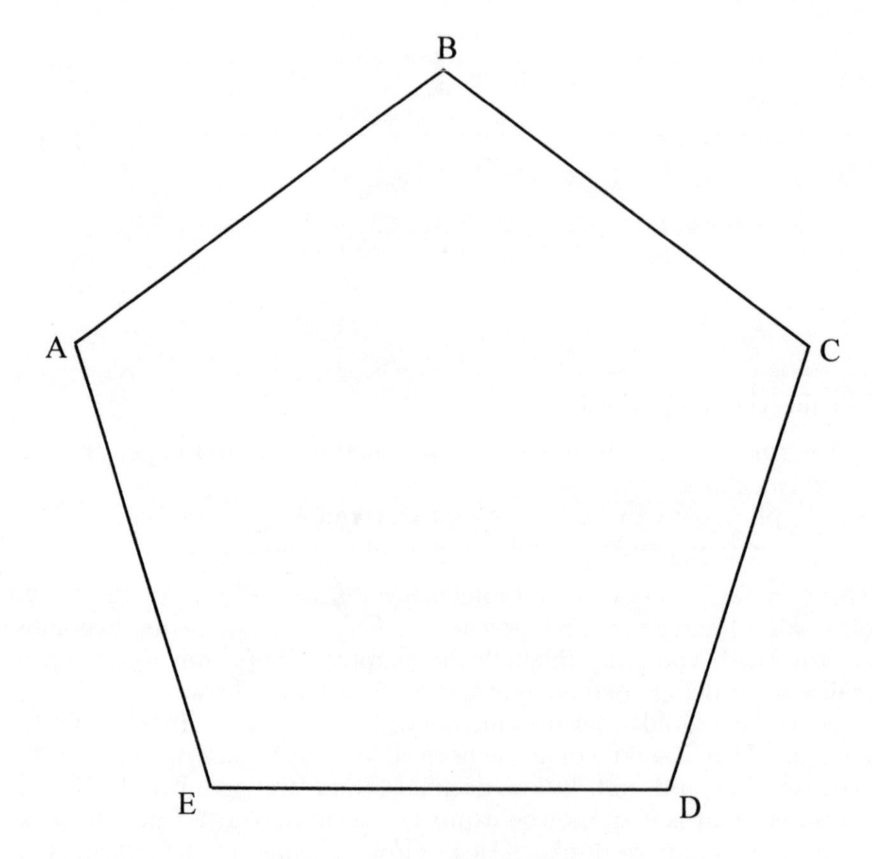

Figure 1: A regular pentagon

A regular pentagon has five sides which are all the same length as each other, and all its angles are the same as each other, too.

1. Draw the line AC on your copy of this regular pentagon. (Draw AC means: draw a straight line from corner A to corner C.)
2. Measure the length of the line AC – in millimetres – and write it down.
3. Measure the length of AB and write it down.
4. Divide the length AC by the length AB; do AC ÷ AB .
5. AC is a diagonal of the pentagon. Draw all the other diagonals: AD, BD, BE and CE.

The five pointed star inside the pentagon is called a pentagram.

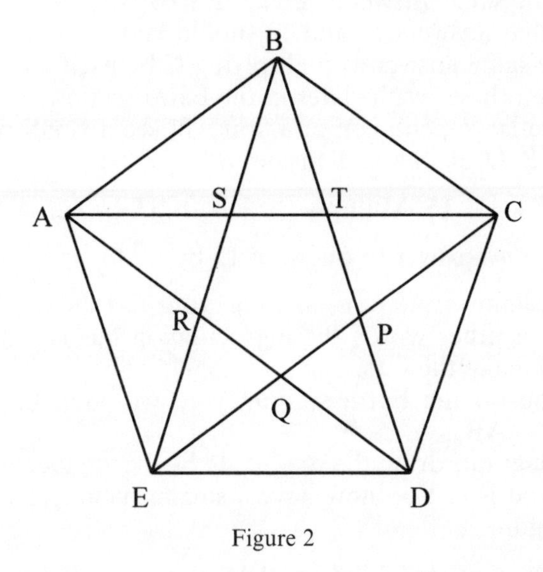

Figure 2

6. On your own diagram, mark the points where the diagonals cross as P, Q, R, S and T as shown in figure 2.
7. Measure the length of AT (in millimetres).
8. Divide the length of AC by the length of AT; do AC ÷ AT.

The answers to questions 4 and 8 should be the same. This number is known as the Golden Section – and also as the Golden Ratio or Divine Proportion. It is one of those numbers which can go on for millions of decimal places, but for our purposes, we will take it as 1.618.

In the Extra section, we will show that it is $\frac{\sqrt{5}+1}{2}$

9. Check with your calculator that this is in fact 1.618 approximately. To do this, find the square root of 5, add 1 then press = to find what the top line comes to; then divide this answer by 2.

From now on, to save space, I will not write 1.618, but the letter *G* instead. (This is algebra – I'm sorry about that!) Every time you see *G*, think of it as 1.618.

Do these calculations: **10**. $G - 1$ **11**. $1 \div G$ (Note $1 \div G = \frac{1}{G}$) **12**. $G + 1$ **13**. G^2 ($G^2 = G \times G$)

Note that answers 10 and 11 should be the same – i.e. taking 1 away from G gives he same answer as dividing 1 by G.

Note also that answers 12 and 13 should be the same – i.e. adding 1 to G gives the same answer as multiplying G by itself.

You may use these results later in the Extra section.

Look at the large pentagon and you will see a small pentagon with its corners at P, Q, R, S and T inside.

14. Measure ST **15**. Multiply the length ST by G (i.e. ST x 1.618).

16. Multiply your answer to question 15 by G.

Allowing for slight errors in measuring, your last answer should be the same as AB, in other words the big pentagon has shrunk by G, then by G again to make the little one.

Note: If you do the Extra section, you will soon be able to show that ST x G^2 = AB.

On your diagram, draw the diagonals of the smaller pentagon: PR, PS, QT, QS and RT. You now have a smaller pentagram, and a much smaller pentagon.

17. Looking at questions 14, 15 and 16 above, see if you can calculate the length of one of the sides of this very small pentagon. Measure to see if you are correct.

Here is another way of finding the Golden Section. It may seem confusing at first, but stay with it!

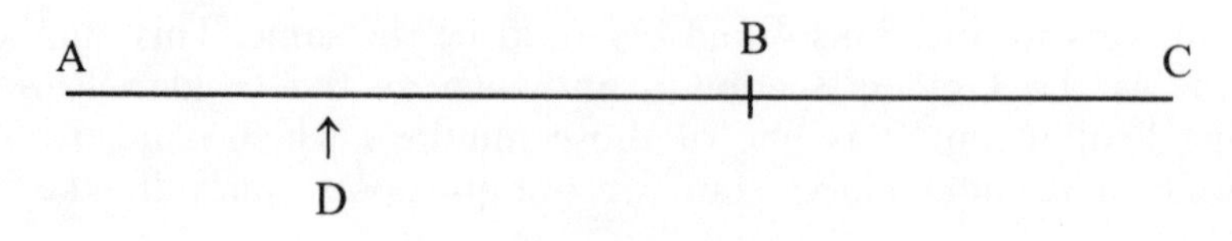

Figure 3

18. Measure AB. **19**. Measure BC.

20. Divide AB by BC (Do AB ÷ BC).

You should get G, approximately.

If you bend BC round through half a turn, C will reach D; in other words, BC = BD.

21. Measure BD. **22**. Measure AD.

23. Divide BD by AD (Do BD ÷ AD).

You should get G again, although because the lengths are shorter, you may well be a little way off 1.618.

Summarising the last six questions: if you cut a line in a certain ratio and the piece you cut off divides the piece you have left in the same ratio, this ratio is the Golden Section.

There are some more clever bits on the Golden Section in the Extra section later. They are not included here in case they frighten some of you off!

Next we go back to some drawing. You will need a photocopy or a tracing of the rectangle, figure 4.

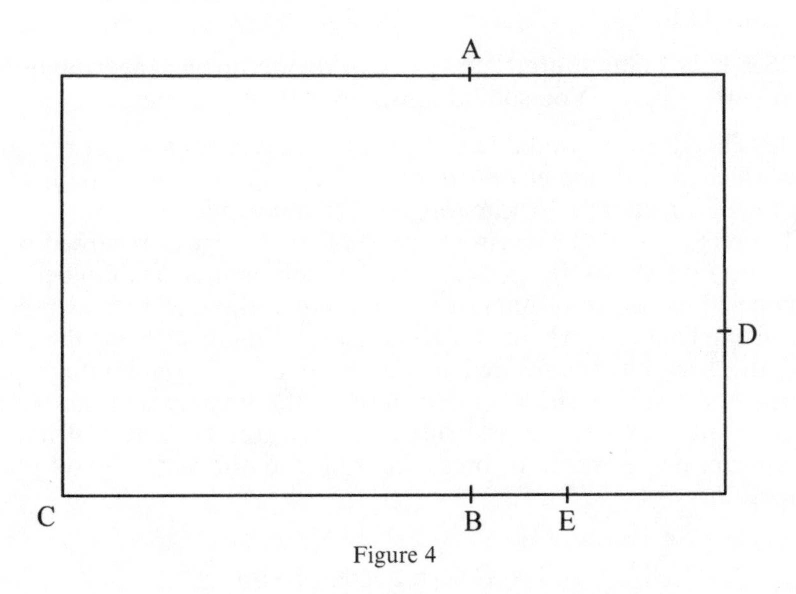

Figure 4

The ratio of the length to the breadth of this rectangle is $G : 1$ (i.e. $1.618 : 1$).

24. Join A to B. This will give you a square and a smaller rectangle.

25. Measure the length and breadth of the smaller rectangle and divide the larger by the smaller to show that they make the Golden Section.

Note: it is obvious if you think about it that these lengths will be in the Golden Section, because you have cut off a shorter length in the Golden Section, so that it and one side of the original rectangle will also be in this ratio.

26. Put a compass point on B and draw a quarter circle from C to A.

27. Draw a line from D across the rectangle to the left, up to where it meets AB.

You will now have another square and an even smaller rectangle (with its sides in the Golden Section, of course!).

28. Put a compass point on the point where the line from D meets the line AB and draw a quarter circle from A to D.

29. Draw a line from E straight up the page until it meets the line from D.

30. Put a compass point where the lines meet and draw a quarter circle from D to E.

31. See how many more times you can divide up rectangles and draw quarter circles. You should finish up with a nice spiral.

Note: This drawing would have been much easier on a larger rectangle, one which would not fit into this book. If you want to draw it again, start with a rectangle 162mm long and 100mm wide.

The rectangle with sides in the ratio of 1.618 : 1 was regarded by the Ancient Greeks to be perfect, hence such words as 'Golden' and 'Divine'. The centre of interest in a picture is thought to be where the line from D meets AB in the rectangle of Figure 4. Here the length and the breadth are divided in the ratio of $G : 1$. Photographers approximate this position to one third of the way up and across, and freely use the expression 'the rule of thirds' for the centre of interest. Of course, it is possible to break the rule and still take a good photograph.

The Golden Section Extra

This section is not for the particularly faint-hearted, but have a go and see!

The next section of work is included because it leads into the Fibonacci Sequence.

Reminder: $G^2 = G \times G$; $G^3 = G \times G \times G$ ($= G^2 \times G$); $G^4 = G \times G \times G \times G$ etc.

$G = G$ (Well, well!)

$G^2 = G \times G = G + 1$ (See earlier in this chapter.)

$G^3 = G \times G \times G = G \times G^2 = G \times (G + 1)$
(Because $G^2 = G + 1$)

$G^3 = G^2 + G$ (So G^3 is the sum of the two previous terms.)

$G^4 = G \times G \times G \times G = G^2 \times G^2 = G^2 \times (G + 1)$ (Because $G^2 = G + 1$)

$G^4 = G^3 + G^2$ (So G^4 is the sum of the two previous terms.)

All this is just to show that the powers of G (G^1, G^2, G^3 etc.) are each the sum of the previous two powers. Using this we get the pattern in a more interesting way:

$G = G$

$G^2 = G + 1$

$G^3 = G + G + 1 = 2G + 1$

$G^4 = G^2 + G^3 = G + 1 + 2G + 1 = 3G + 2$

$G^5 = G^3 + G^4 = 2G + 1 + 3G + 2 = 5G + 3$

32. Develop this sequence further if you have the courage; find G^6, G^7 etc.

33. Check any of these results on your calculator, remembering that $G = 1.618$ approximately.

The next section is definitely not for the faint-hearted! It uses the algebra of quadratic equations to find the value of G.

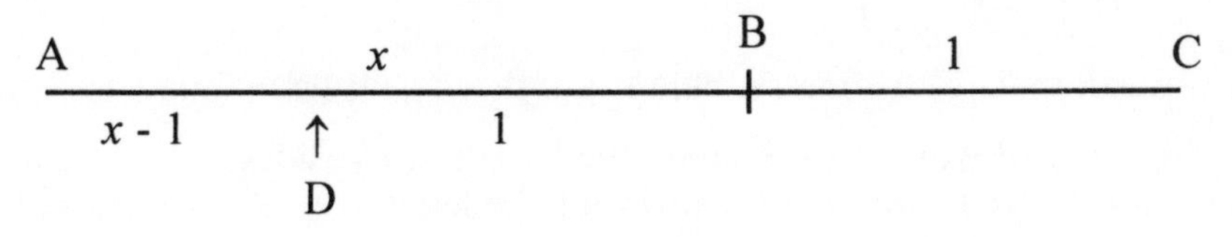

Figure 5

If AB $= x$ and BC $= 1$, then BD $= 1$ and DA $= x - 1$.

The ratio AB : BC is the same as the ratio BD : DA .

This is : $x : 1 = 1 : (x - 1)$

Writing these as fractions: $\frac{x}{1} = \frac{1}{(x-1)}$

So, cross-multiplying: $x(x - 1) = 1 \times 1$

Expanding the bracket and rearranging: $x^2 - x - 1 = 0$

The formula for solving $ax^2 + bx + c = 0$ is $x = \frac{-b \pm \sqrt{(b^2 - 4ac)}}{2a}$

In our equation: a = 1 ; b = –1 ; c = –1 .

Substituting in the formula: $x = \frac{1 \pm \sqrt{[1 - 4 \times 1 \times (-1)]}}{2} = \frac{1 + \sqrt{(1+4)}}{2}$

Ignoring the negative root:

$$x = \frac{1 + \sqrt{5}}{2} \quad \text{or} \quad x = \frac{\sqrt{5} + 1}{2} \quad = G$$

We can now show two more results:

$$G^2 = G \times G \quad = \frac{(\sqrt{5} + 1)}{2} \times \frac{(\sqrt{5} + 1)}{2} = \frac{5 + 2\sqrt{5} + 1}{4}$$

$$= \frac{6 + 2\sqrt{5}}{4} \quad = \quad \frac{3 + \sqrt{5}}{2} \quad = \quad \frac{\sqrt{5} + 1}{2} + 1$$

$$= G + 1$$

And:

$$\frac{1}{G} \quad = \quad \frac{2}{\sqrt{5} + 1} \quad = \quad \frac{2(\sqrt{5} - 1)}{(\sqrt{5} + 1)(\sqrt{5} - 1)} \quad = \quad \frac{2(\sqrt{5} - 1)}{5 - 1}$$

$$= \quad \frac{\sqrt{5} - 1}{2} \quad = \quad \frac{\sqrt{5} + 1}{2} - 1 \quad = \quad G - 1$$

If you can manage geometry, here is another result from figure 1:

In this piece of work, I have taken the length of the sides of the regular pentagon to be 1 unit. This means that the length of AC, for example, is 1.618... or G.

AC = G and AT = 1, so because TC = AC – AT then TC = G – 1.

Also BT = G – 1 because triangle BTC is isosceles; its base angles at B and C are equal.

BD = G and triangles BTS and BDE are similar.

Because BT = $\frac{1}{G}$(= G – 1) the ratio of corresponding sides is $G : \frac{1}{G}$

This is much better written as $G^2 : 1$, which is the ratio of ED to ST.

That is why the large pentagon is G^2 times bigger than the smaller one.

Phew!

2

The Fibonacci Sequence

For this chapter you will need:

- Pen and paper
- Scissors and glue
- Photocopies or tracings of figures 1, 2 and 3
- A calculator
- A computer if you have one

The Fibonacci Sequence is easy to generate, but it seems to have endless fascination for mathematicians. I have included just a few of the many facts which could have been included.

Mathematicians and scientists have searched for centuries for patterns in nature. This is touched upon in this chapter when you make three rather stubby trees. However, you will see that the tree from figure 1 is most unstable; I would not like to be under it in a high wind. The tree from figure 2 is more stable, but not much like any tree I know. The tree from figure 3, related to the Fibonacci Sequence – also to the Golden Section – has its branches 'growing' from the trunk in a much more familiar style. Does mathematics reveal a pattern in nature here?

The Extra section is for those of you who have a computer with a spreadsheet program. Try it, even if you have never used a spreadsheet before. A computer can generate then graph the Fibonacci Sequence with ease. The graph of the division of successive terms gradually reaching a limit is the best way I can think of to demonstrate this mathematical point.

Vocabulary: A sequence in mathematics is just a string of numbers, usually with a pattern to it. The numbers are referred to as 'terms' of the sequence such as: 'the first term of the sequence' (sometimes T_1), 'the second term of the sequence' (T_2) etc.

1. Here are the first five terms of the Fibonacci Sequence. Copy them, find out how you get each term from the previous two, then continue it until you get fed up!

1, 1, 2, 3, 5, ...

2. Now divide each term by the one before it – until your patience runs out.

e.g. $1 \div 1 =$ $2 \div 1 =$ $3 \div 2 =$ $5 \div 3 =$ etc.

This of course makes another sequence, and is an example of a sequence which approaches a limit. If you get 10 or more terms, you may see what the limit is.

3. When you add two odd numbers together, is your answer odd or even?

4. When you add two even numbers together, is your answer odd or even?

5. When you add an odd number to an even number, is your answer odd or even?

6. Work out the pattern for odd (O) and even (E) numbers in the Fibonacci Sequence; it has been started for you:

1 1 2 3 5 ...
O O E O O ...

7. Use your answers to questions 3, 4 and 5, together with your knowledge of the Sequence, to work out why you get this pattern.

We can summarise questions 6 and 7 by saying that every third term is divisible by 2.

8. What number divides exactly into every fourth term?

9. What number divides exactly into every fifth term?

10. What number divides exactly into every sixth term?

11. Can you see a pattern in the answers to the last few questions?

Fibonacci (meaning 'son of good fortune'?) lived in the 13th century. His real name was Leonardo of Pisa and he worked for his father, a wealthy merchant who traded among the Mediterranean ports. He wrote a book called *Liber Abaci* or 'book of calculations'. He was nicknamed Fibonacci by Edouard Lucas, of whom more later.

Mathematicians are so keen on finding odd facts about the Fibonacci Sequence that there is a journal in the USA called the

Fibonacci Quarterly which is full of obscure findings. Here are some I have found:

There is no simple way to find, say, the tenth term of the sequence without going all the way through it. However, here is a simple rule which seems to work; try it and see:

12. (In maths speak) T_n is the nearest whole number to $G^n \div \sqrt{5}$, where $G = 1.618$.
(In English) To find, say, the tenth term, multiply G by itself 10 times then divide your answer by $\sqrt{5}$ (which is 2.236); then take the nearest whole number. Check the calculation to see if this is true.

13. Here is another cheerful fact: try the calculation for various terms and see if you can spot the pattern of the results.
(In maths speak) Find $T_{n-1} \times T_{n+1} - T_n \times T_n$
(In English) Multiply the first term by the third term, then take away the second term multiplied by itself.
Multiply the second term by the fourth term, then take away the third term multiplied by itself, etc. etc.
Check a calculation to see what you get.

Here are some examples of where the terms of the Fibonacci Sequence can be found all around us. Sometimes this is by chance, but not always.

Music:

1. A simple piece of music is just that; one piece of music!
2. Clementi and others of his time wrote piano sonata movements which were in two sections.
3. Classical sonata form then developed into three sections: exposition, development and recapitulation.
5. Bartok developed an arch form in five sections: ABCB′A′.
   On a piano, within an octave from C there are 5 black notes.
8. In the same octave, there are 8 white notes.
   There are, of course, 8 notes to a major scale.
13. A chromatic scale on the piano touches 13 keys, the 5 black and the 8 white mentioned above.

Botany:

Frank Land in *The Language of Mathematics* quotes:

2. Two-petalled flowers are rare. One example is the Enchanter's Nightshade.
3. Three-petalled flowers are found among lilies and irises.
5. Five-petalled flowers are the most common.

Multi-petalled flowers such as daisies and dandelions can have petals numbering larger terms of the Fibonacci Sequence. The classic example of Fibonacci numbers is in the number of spirals and the number of seeds in a spiral of the seed-head of the sunflower. My eyesight is not good enough to see if this is also true of the common daisy, but it probably is. Another example closer to home is to look at the pattern of prongs on a pine cone.

Phyllotaxis is the study of the arrangement of leaves growing from a stem.

David Wells in *Curious and Interesting Numbers* reports on the arrangement of leaves growing from a stem. Each angle referred to below is the angle you see between successive leaves if you look down on the stem from above.

The commonest angles are: 180°, 120°, 144°, 135° and approximately 137.5°.

180° divides the circle in the ratio 1 : 2
120° divides the circle in the ratio 1 : 3
144° divides the circle in the ratio 2 : 5
135° divides the circle in the ratio 3 : 8
137.5° (approx) divides the circle in the ratio 1 : 1.618 or 1 : G (the Golden Section)

This last angle is sometimes referred to as the Golden Angle.

An angle of 180° would mean that any leaf would be directly under the one which is two places above it on the stem.

An angle of 120° would mean that a leaf would be directly under the one three places above it on the stem.

The Golden Angle would mean that no leaf would be directly under another.

This gives a good spread to the leaves of a bush and increases the amount of daylight each leaf receives.

Mathematicians say that a number which can be written as a fraction is 'rational' (can be written as a ratio) whereas a number which cannot, such as G, is 'irrational'.

$$G = 1.618034 \ldots \text{(approx!)}$$

Here is a piece of nonsense to illustrate the work on phyllotaxis.

Figures 1, 2 and 3 at the end of this chapter are diagrams for you to copy, cut out then stick together. You will have a 'tree' from each sheet, and six 'branches'. Cut out each 'tree' then pierce where the holes are marked. Roll it into a cylinder before gluing the tab to its opposite side. Cut the tabs at the bottom of the tree and bend them out so that the tree can stand up. Cut out the six branches then roll them into cylinders and push them into the holes in the tree.

Note how, in the first two 'trees', leaves at the top of the tree cover those lower down. Try blowing them over to see which tree is the most stable.

The Fibonacci Sequence Extra

Computer Use:

If you have a computer you can use it to generate the Fibonacci Sequence, then draw its graph. It will also show the division of successive terms tending to a limit. The instructions below work on my spreadsheets; I hope they work on yours!

Load your spreadsheet.

Note that each row is numbered 1, 2, etc. and that each column is lettered A, B, C, etc. Therefore the cell (the rectangular space) in column C and row 5 is cell C5 (or c5).

Point the mouse to cell B1 and click. Type 'Term 1'. Now move the mouse pointer to the bottom right-hand corner of cell B1; a cross should appear. Drag this cross (i.e. move the mouse while holding down the left-hand button) to the right about 26 cells. You should see that the cells automatically number 'Term 2', 'Term 3' etc.

For the rest of this chapter, I will assume that you dragged to cell Z1.

Go back to cell A2, using the slider at the bottom of the screen, and in it type 'Fibonacci Number'. It may not all show, but it will do so in the final graph.

In cell B2, enter the number 1. Enter the number 1 in C2 also. These start off the sequence. You can change these numbers later, but do not leave either cell empty, or enter 0.

Click on cell D2 then enter '=B2+C2' (or '=b2+c2'). This is a formula which will put into cell D2 the sum of the numbers you have put in cells B2 and C2. If later you change the numbers in B2 or C2, the total in D2, and subsequent cells, will change automatically.

Move the mouse pointer to the bottom right-hand corner of cell D2 and drag it to the right as you did when labelling the cells above. Stop where you finished labelling the row above.

If you go back (with the slider) and click on cell E2, you should see towards the top of the page in the formula window the formula '=C2+D2', and in cell F2, there will be the formula '=D2+E2'. The spreadsheet has automatically changed the formula for the cells across to the right. The numbers in the cells are the numbers of the Fibonacci Sequence, taken as far as you dragged the mouse.

Now to draw the graph:

Place the mouse pointer over cell A1 and drag it (move it while holding down the left hand button) to cell Z2. All the cells should now be selected; you will see that the printing is white on black, not black on white.

Now click 'Tools' at the top of the screen, then click on 'New Chart'. You should now be able to choose which kind of graph the computer will draw for you.

Note that you can save your data to hard or floppy disk.

Go back to the spreadsheet for more adventures!

Click on cell C1, and type 'Golden Section'.

In cell C3, enter '=C2/B2'. This will divide the contents of cell C2 by the contents of B2. It is because you cannot divide by zero that there must be something other than 0 in each cell. If there is no number entered into a cell, the computer assumes that the number 0 has been entered.

Point to the bottom right-hand corner of cell C3, provided it is still selected, and drag across to the right, again as far as you have entries in the row above.

Select all the cells from A1 to Z3 by dragging the mouse pointer, then go to 'Tools' and 'New Chart' again.

You may get a better view of the divisions tending to *G* by selecting row 3 on its own.

The graph of Row 3 should show that the graph 'wobbles' a bit at first, and then settles to a straight line. If you look at the cells around Z3, all the values of *G* will be the same.

The French mathematical writer Edouard Lucas, who gave Leonardo of Pisa the name Fibonacci, experimented with different starting numbers to the sequence. He had to use pen and ink, but you can use a computer by changing the entries in cells B2 and C2, but remember to put something in them, and something other than 0.

The first two numbers of what are called Lucas Numbers are 1 and 3.

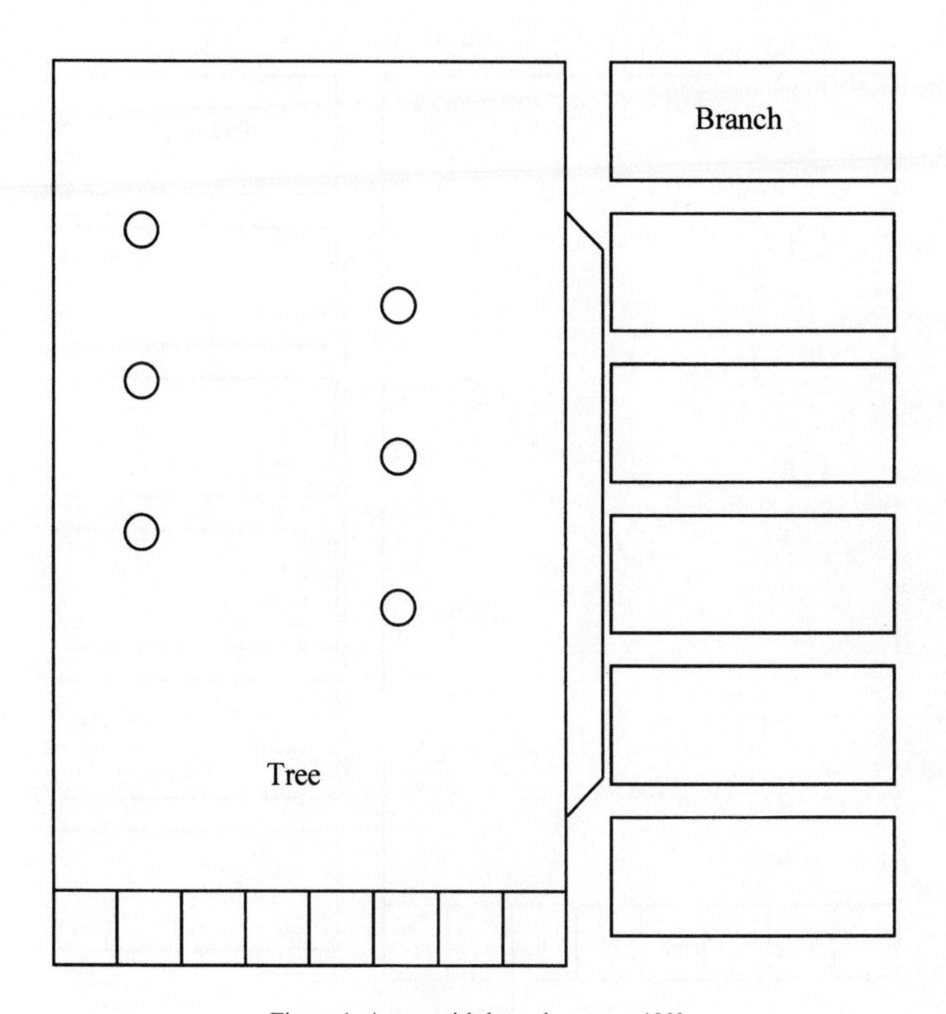

Figure 1: A tree with branches every 180°

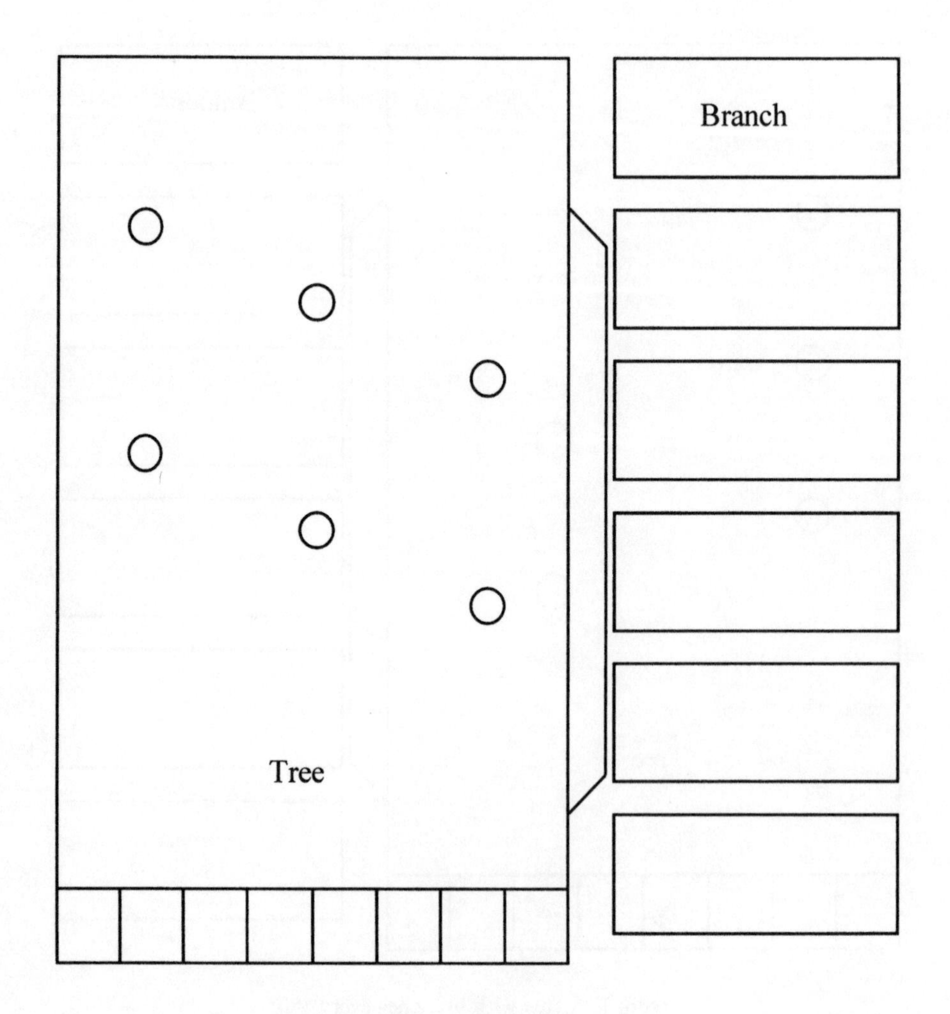

Figure 2: A tree with branches every 120°

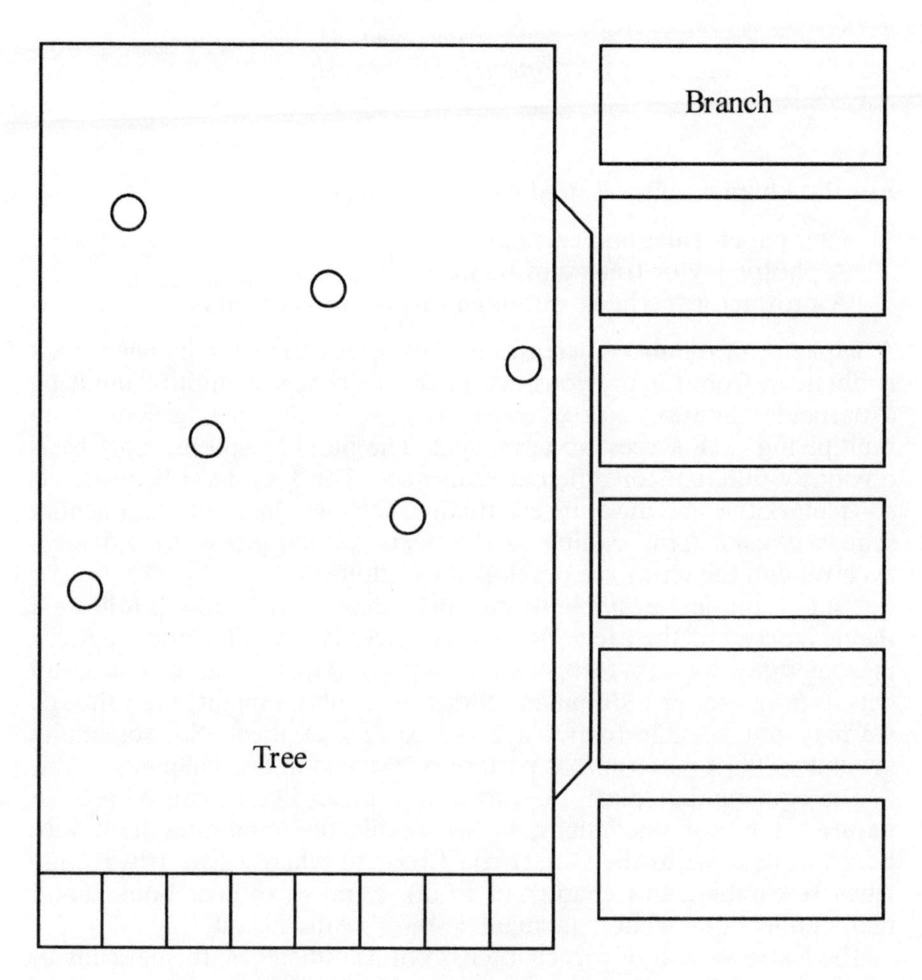

Figure 3: A tree with branches every 137.5°

3

Exponential Growth

For this chapter you will need:

Pen, paper, ruler and calculator
A photocopy or tracing of figure 1
A protractor, perhaps, although this is not necessary

A sequence of numbers increases or decreases exponentially when each is obtained from the previous one in the sequence by multiplying it by a particular number. For example: 1, 2, 4, 8, 16, ... is generated by multiplying each successive term by 2. The number you multiply by is obviously different for different sequences. The Fibonacci Sequence is so remarkable because it eventually behaves like an exponential sequence, each term leading to the next by multiplying by 1.618 ... even though the terms are developed by addition.

Using a musical example to introduce exponential growth follows a theme (sorry!) of the previous chapter: there is a mathematical pattern in something we can recognise around us. Most of us can hear an octave in music, or a dominant chord or a subdominant, even though we may not be able to put a name to any of them. So, something around us fits a pattern; that pattern is revealed in this chapter.

The equiangular spiral, or spirals very much like it, can be seen in nature. It is not uncommon to see fossils of ammonites built into house walls close to the Dorset coast near to where I live. It was only when researching this chapter in Frank Land's excellent book that I really understood what 'equiangular spiral' really meant.

The Extra section of this chapter is not as difficult mathematically as some of the others. Have a look at it and see!

Here is a note which may help you with question 9: if a sequence of numbers is, say, 64, 16, 4, 1, ... mathematicians are quite happy to say that you get each term by multiplying the previous one by something. In this case, that something is ¼. Most normal people would say that you divide by 4, but are mathematicians normal people?

When an orchestra tunes up, one instrument, usually an oboe, plays a note and the other instruments tune relative to it. That note is the A above middle C which is the note which sounds when air vibrates at 440 beats per second. (Some take it as 441!) A note one octave (8 notes or 13 semitones) below this is heard when air vibrates at 220 beats per second. (220 is half of 440!) An octave above the A will beat at 880 beats per second. The 12 spaces between the 13 semitones of a scale are equally divided these days. This division is called 'equal temperament' and is what J.S. Bach meant when he used the term 'The Well Tempered Clavier'.

The technical term for 'beats per second' is 'Hertz'; A is 440 Hertz.

As each note rises in pitch by one semitone, the number of beats per second increases by 1.0595 times.

1. Copy and finish this table which has been started.
   (Note: If you have a calculator with an auto-constant, you can take the increase in Hertz to be 1.0594631)

Note	Hertz	... to the nearest whole number
A	220	220
A#	220 × 1.0595 = 233.09	233
B	233 × 1.0595 = 246.96	247
C	247 × 1.0595 =	
C#		
D		
D#		
E		
F		
F#		
G		
G#		
A		

2. When playing music in the key of A, the key you are most likely to drift into is E, which is why it is called the dominant key of A. Allowing for slight errors in approximating your answers, what fraction of the way up the scale – in Hertz – is E?

3. How many semitones is E above A?

4. Which note is the same number of semitones *below* A?

5. In the key of D, A is the dominant. In the key of A, D is called the subdominant. Very approximately, what fraction of the way up the scale is D?

The sequence $\frac{1}{2}$ $\frac{1}{3}$ $\frac{1}{4}$ $\frac{1}{5}$ etc. is called a harmonic sequence.

When a quantity increases by the same (multiplied) amount, this is called exponential growth. On some calculators, when a question has an answer such as 1234567890 and it is too big for the display, it will be shown as 1.234567 E9. This means 1.234567×10^9 and the E stands for 'exponent'.

There is a picturesque way of showing information like that of the table above which is there for you to copy from figure 1. It takes patience and care but it is rewarding. Follow these instructions:

Turn the sheet through a right angle so that the words 'Start here' are on the right.

The numbers in the table on the right are in two columns, going downwards, and show lengths in millimetres.

Place your ruler along the line to the right, towards where it says 'Start here'.

Move it so that the end of its measure is where all the lines cross.

Put a small dot 8.3mm along this line. (The first number in the table is 8.3.) You cannot expect to measure as accurately as a decimal place of a millimetre, but the figures are put in this way in case you want to explore the pattern of the table.

Move anti-clockwise to the next radial line and put a dot as best you can 8.8mm from the point where all the lines cross. (The second number in the table is 8.8.)

Move anti-clockwise and put a dot 9.3mm from the centre, then keep going round and round until you run out of numbers.

Either as you go, or when you have finished, join the dots you make in the order you made them; you should get a spiral. If you are hopeless like me, join each pair of dots with a straight line, or if you are clever, join them with a smooth curve.

It might help you keep your place if you cross out each number in the table as you use it.

The numbers could have started below 8.3, the lowest number here, but it would have been almost impossible to have drawn them with any degree of accuracy.

6. If you can, measure the angle between each line you have just drawn (joining the dots) and the nearest radial line. It should be the same for each radial line. (For mathematicians, this should be 'measure the angle between the radial line and the tangent to the curve'.)

All these angles should be the same, which is why this curve is known as an 'equiangular spiral'.

You should be able to find equiangular spirals in the patterns on sunflowers, daisies and pine cones.

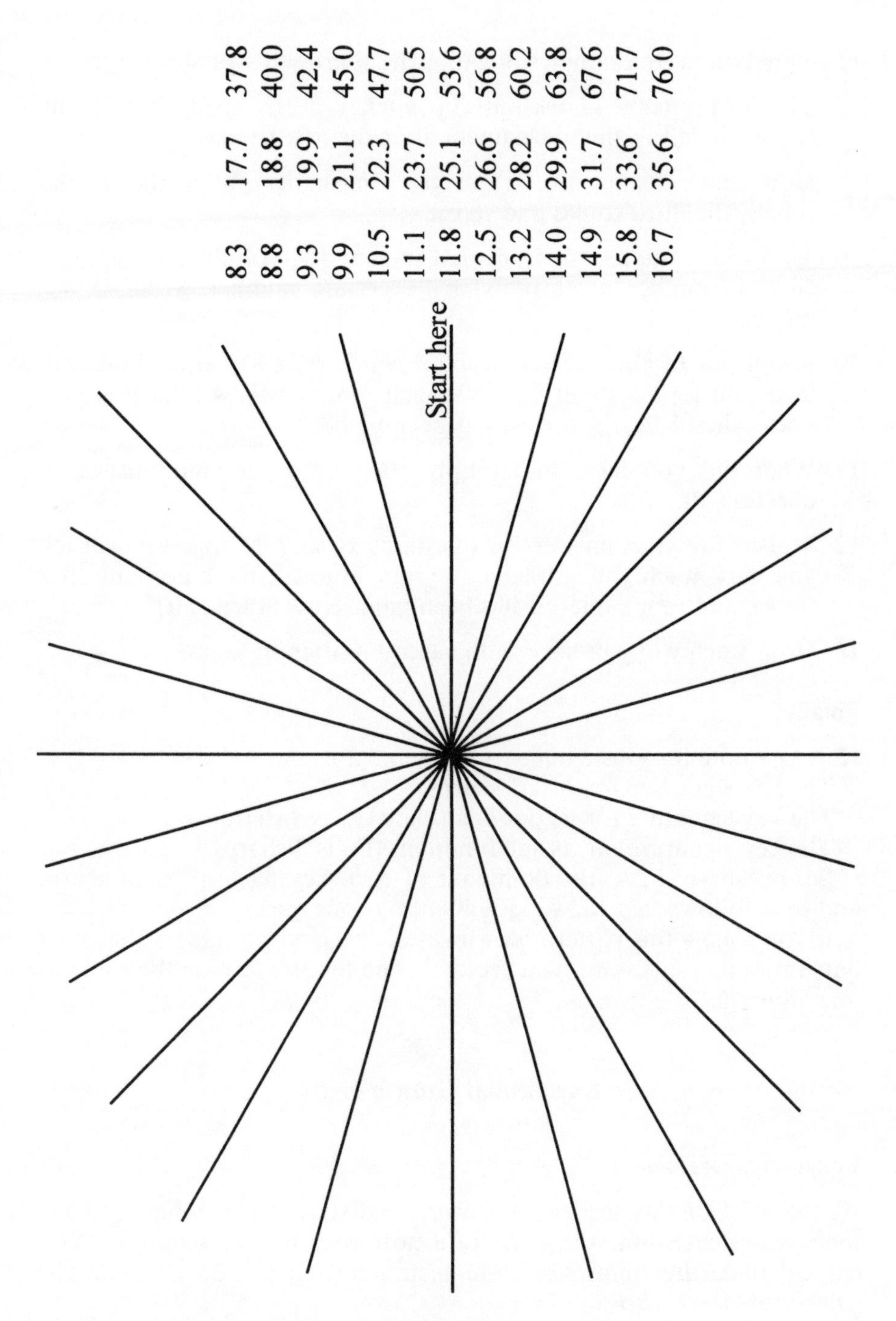
8.3 17.7 37.8
8.8 18.8 40.0
9.3 19.9 42.4
9.9 21.1 45.0
10.5 22.3 47.7
11.1 23.7 50.5
11.8 25.1 53.6
12.5 26.6 56.8
13.2 28.2 60.2
14.0 29.9 63.8
14.9 31.7 67.6
15.8 33.6 71.7
16.7 35.6 76.0
Start here

Figure 1

Here are two more examples of exponential growth – or shrinkage!

7. 64 teams enter a knock-out cup hockey tournament. How many teams are left in the tournament after the first round?

8. How many teams are left in the tournament after the second round, the third round and so on.

9. To find the number of teams left in round 2, you have to multiply 64 (the number of entrants) by a certain number. What is this number?

10. If you put £100 in a bank account which pays 8% annual interest, then you forget about it, how much money will you have after 1 year (when 1 year's interest has been paid)?

11. What did you have to multiply 100 by to get your answer to question 10?

12. Multipying your answers to questions 10 and 11 together will tell you how much you will have in your forgotten bank account after the second year's interest has been paid. How much is it?

13. How much will you have in the account after 10 years?

Finally:

Here is a note for music buffs (pun intended!).

The key signature for A is 3 sharps.

The key signature for its dominant key (E) is 4 sharps.

The key signature for its subdominant (D) is 2 sharps.

So, in 'sharp' keys, the dominant of each key has one extra sharp, and so it follows that the subdominant has one fewer.

If you follow this pattern 'downwards' to C, which has no sharps or flats in its key signature, you see that 'one fewer sharp' is the same as 'one more flat'.

Exponential Growth Extra

Equal Temperament 1

At the start of this section, you were told that strings vibrate 1.0595 times more each time the pitch of a note rises by one semitone. You can calculate this number – more accurately – if you have a scientific calculator. Here's how:

To go up an octave the number of Hertz must double (× 2). To go up one semitone, or one twelfth of an octave, the number of Hertz must increase by the twelfth root of 2. To find this, take the logarithm of 2, divide it by 12, then anti-log – like this:

Press 2; log; (newer calculators log then 2); ÷; 12; =; 10^x.

Equal Temperament 2

On instruments where the notes are not fixed for the player, such as violins and cellos, it is possible to play music using a scale which is not equally tempered. It is possible to hear the difference between an equal tempered scale and a 'natural' one.

Equiangular Spiral

You will have noticed that the important thing about exponential growth is that something – a length, Hertz, money – grows or shrinks by the same multiplied amount. This amount is often called the scale factor; in money it is sometimes called the growth factor or simply the multiplier. The scale factor for the equiangular spiral was 1.06; this means that the lengths were increased by 1.06 each time you moved one line anti-clockwise. It is close to the factor for equal temperament but that is coincidence; it was chosen so that a good spiral would fit on to the page.

The characteristic angle between the spiral and each radial line is 77.5° for this particular spiral.

There are 24 radial lines in one turn of this spiral, so if the spiral cuts a particular radial line twice, the second distance is 1.06^{24} times further from the centre than the first. $1.06^{24} = 4.05$ approximately. For example, the fourth point is 9.9mm from the centre, and the 28th point $(28 = 24 + 4)$ is 40mm. $9.9 \times 4.05 = 40$ to the nearest whole number. Check some others for yourself.

4

Fractions

For this chapter you will need:

Pen and paper
A calculator if you wish to use one

For much of this book, I have tried to tie what you do into some aspect of the real world. The next piece of work is different: there is no particular application to the work of this chapter other than improving your arithmetic if it is rusty.

If you removed from human experience everything which had no particular use, there would not be much left for us to do. Mathematicians are particularly fond of studying something just for the sake of it. So noble do they think this is that they refer to it as 'pure mathematics', as though maths with an application is something to be avoided!

However, there is one important principle of mathematical thinking here, and that is pattern. In this chapter, we look for any pattern there is in numbers, which I always think is one step further in mathematical thinking than looking for a pattern in shapes; it certainly is for young children.

Question 6 may challenge what you know about recurring decimals, particularly accepting that 0.9999999... is 1, or at least we can never measure how wrong we are if we choose to doubt it.

Another principle which is developed a little is the idea of a sequence of cyclic numbers. We are all familiar with things cyclic. If your week starts on a Monday, when you get to the next Sunday, you call the day following it 'Monday'. The pattern has gone round one cycle.

Times, which are measured in seconds, minutes, hours and seasons, are all cyclic and for good reason. When it is winter, I like to think that is going to be spring before long. It always is!

The word 'vulgar' means 'common', so vulgar fractions are just common ones. This exercise is to see what happens when vulgar fractions are converted into decimal fractions. From now on, decimal fractions will be called 'decimals' – it's easier! Although there will be times when a calculator is useful, it often obscures the patterns which emerge from investigating this exercise.

You only need fractions when you divide something. For example, if you have half a cake, you have taken 1 – the whole cake – divided into 2 equal parts. The divide symbol ÷ is a fraction with a dot above and below the line in place of numbers. So to change the fraction $\frac{1}{2}$ into a decimal, it becomes

$$2\overline{)1.0} \quad \text{which is } 0.5$$

1. By 'hand' – not with a calculator – change $\frac{1}{3}$ and $\frac{2}{3}$ into decimals, that is

$$\text{do} \quad 3\overline{)1.000} \quad \text{and} \quad 3\overline{)2.000}$$

2. This question is an investigation – more precisely a closed investigation. I set up a problem and you answer it! It is 'closed' because there is a definite end to it.
   Investigate what happens when $\frac{1}{7}$, $\frac{2}{7}$, $\frac{3}{7}$, $\frac{4}{7}$, $\frac{5}{7}$, and $\frac{6}{7}$ are each changed to decimals. The first one has been done as an example

   $\frac{1}{7}$ is $1 \div 7$ which is

$$\begin{array}{r} 0.\,1\;4\;2\;8\;5\;7\;1\ldots \\ 7\overline{)1.{}^{1}0{}^{3}0{}^{2}0{}^{6}0{}^{4}0{}^{5}0{}^{1}0} \end{array}$$

   Notice that the division is back where it started, so the pattern of numbers will repeat. Now it's your turn. Finish the investigation, but if you have sharp eyes, you will not need to do 5 more divisions.

3. Now investigate the ninths; this time a calculator could be used. Change these to decimals: $\frac{1}{9}$ $\frac{2}{9}$ $\frac{3}{9}$ $\frac{4}{9}$ $\frac{5}{9}$ $\frac{6}{9}$ $\frac{7}{9}$ $\frac{8}{9}$

4. You have the answer to $\frac{1}{3}$ as a decimal. How many other thirds are included in the ninths?

5. Investigate the elevenths in the same way as question 3; a calculator would be useful.

6. The fraction $\frac{1}{3}$ is 0.333333 … and $\frac{1}{5}$ is 0.2. $\frac{1}{15} = \frac{1}{3} \times \frac{1}{5}$

   $\frac{1}{15}$ as a decimal is $0.33333 \ldots \times 0.2$

   (a) Multiply the decimal fractions together to find $\frac{1}{15}$ as a decimal.

   (b) $\frac{2}{15}$ is obviously twice $\frac{1}{15}$. Double your answer to (a) to find $\frac{2}{15}$

(c) $\frac{3}{15}$ is $\frac{1}{15} + \frac{2}{15}$. Add your answers to parts (a) and (b) to find $\frac{3}{15}$

Notice here that 0.2 is the same as 0.19999..., isn't it?

7. Here is the last example of converting a vulgar fraction to a decimal. It is long and tedious, but is included because of some work later on. Change $\frac{1}{17}$ to a decimal. This will have to be done the long way, so to help, here is the 17 times table:

$1 \times 17 = 17$ $2 \times 17 = 34$ $3 \times 17 = 51$ $4 \times 17 = 68$
$5 \times 17 = 85$ $6 \times 17 = 102$ $7 \times 17 = 119$ $8 \times 17 = 136$
$9 \times 17 = 153$

Stop if you have more than 16 decimal places.

8. Can you see why the 7ths have 6 decimal places before they recur, and the 17ths have 16? Look at the remainders each time you divide.

We have been looking at recurring decimals. To show that decimals recur we sometimes write dots after the numbers, such as 0.234234... but more often we write a dot above the first and last digit of the recurring group. For example:

0.33333... is written $0.\dot{3}$ and 0.234234234... is written $0.\dot{2}3\dot{4}$

What do you usually do after you've been somewhere? Come back! So, now you can turn vulgar fractions into decimals, recurring or otherwise, you must now find out how to turn recurring decimals back into vulgar fractions. In fact, you already know.

9. Cancel each of these fractions: $\frac{27}{99}$ $\frac{3}{9}$ $\frac{81}{99}$ $\frac{666}{999}$

10. What is each of your answers to question 9 when converted to a decimal?

11. Here is a very easy question: cancel the fraction $\frac{142857}{999999}$
Look at question 2 for a hint.

If you have not yet got there, to convert a recurring decimal to a vulgar fraction, make the recurring group the numerator (top) of the fraction, and give the denominator (bottom) as many 9s as there are recurring digits. The next three questions will let you check this.

Using a calculator, change each of these vulgar fractions to decimals:

12. $\frac{123}{999}$ **13**. $\frac{82}{99}$ **14**. $\frac{6281}{9999}$

Fractions Extra

Here is a simple proof of why the rule for changing recurring decimals to vulgar fractions works.

You first have to realise that if you lose two (or a few more) digits from the end of a recurring decimal, it makes no difference to it. This is because it has an infinite number of digits.

Let's change 0.37373737... to a fraction, and let 0.373737... be x, just to be like mathematicians for a while.

	$x = 0.37373737373737\ldots$	equation A
Multiply by 100:	$100\,x = 37.37373737373737\ldots$	equation B
Take A from B:	$99\,x = 37$	
	$x = \frac{37}{99}$	

Cyclic Numbers

Much of the work in this section is taken from Chapter 10 of *Mathematical Circus* by Martin Gardner.

15. With an eye on your answers to question 2, do each of these, with a calculator if necessary:

$1 \times 142857 =$ $\quad 2 \times 142857 =$ $\quad 3 \times 142857 =$

$4 \times 142857 =$ $\quad 5 \times 142857 =$ $\quad 6 \times 142857 =$

Numbers such as 142857 which behave in this way are called cyclic numbers. If you imagine the digits 1, 4, 2, 8, 5 and 7 on a strip of paper pasted around a jam jar, you can see how to get each answer, getting bigger as you multiply by bigger numbers, by starting at a different point on the jam jar and going once round.

16. Looking at your answer to question 7, if you managed to do it, write down – but do not calculate – the answer to:

(0)588235294117647 × 2 =

17. Cancel (i.e. reduce to its lowest terms) this fraction, but do not spend more than a minute trying to do it!

$$\frac{1764705882352941}{9999999999999999} =$$

You may notice that you get a cyclic number when you divide 1 by a certain prime number, then ignore the decimal point. The numbers less than 100 which will generate cyclic numbers in this way are 7, 17, 19, 23, 29, 47, 59, 61, 97.

One William Shanks discovered in the 19th century that dividing 1 by

17389 generated a cyclic number with 17388 digits. Remember that he had no electronic calculators or computers, but obviously plenty of time!

18. Add the first half of 142857 to its second half:

$$\begin{array}{r} 142 \\ +\ \underline{857} \end{array}$$

19. Add the first half of (0)588235294117647 to its second half.

20. Do the same for (0)52631578947368421, which is the cyclic number generated by dividing 1 by 19.

21. Add the digits of 142857 (i.e. do 1 + 4 + 2 + 8 + 5 + 7)

22. Add the digits of (0)588235294117647

23. Add the digits of (0)52631578947368421

24. Remembering how recurring decimals convert to vulgar fractions, how could you predict that the answers to questions 21, 22 and 23 would be in the 9 times table?

25. Below is a huge magic square based on all the arrangements of the cyclic number of questions 20 and 23. It appeared in a 1917 publication *Magic Squares and Cubes* by W.S. Andrews. If you add each row, each column and the two diagonals, you get the same answer of 81. Check some rows, columns and diagonals to see if this is correct.

0	5	2	6	3	1	5	7	8	9	4	7	3	6	8	4	2	1
1	0	5	2	6	3	1	5	7	8	9	4	7	3	6	8	4	2
1	5	7	8	9	4	7	3	6	8	4	2	1	0	5	2	6	3
2	1	0	5	2	6	3	1	5	7	8	9	4	7	3	6	8	4
2	6	3	1	5	7	8	9	4	7	3	6	8	4	2	1	0	5
3	1	5	7	8	9	4	7	3	6	8	4	2	1	0	5	2	6
3	6	8	4	2	1	0	5	2	6	3	1	5	7	8	9	4	7
4	2	1	0	5	2	6	3	1	5	7	8	9	4	7	3	6	8
4	7	3	6	8	4	2	1	0	5	2	6	3	1	5	7	8	9
5	2	6	3	1	5	7	8	9	4	7	3	6	8	4	2	1	0
5	7	8	9	4	7	3	6	8	4	2	1	0	5	2	6	3	1
6	3	1	5	7	8	9	4	7	3	6	8	4	2	1	0	5	2
6	8	4	2	1	0	5	2	6	3	1	5	7	8	9	4	7	3
7	3	6	8	4	2	1	0	5	2	6	3	1	5	7	8	9	4
7	8	9	4	7	3	6	8	4	2	1	0	5	2	6	3	1	5
8	4	2	1	0	5	2	6	3	1	5	7	8	9	4	7	3	6
8	9	4	7	3	6	8	4	2	1	0	5	2	6	3	1	5	7
9	4	7	3	6	8	4	2	1	0	5	2	6	3	1	5	7	8

26. It is obvious that each row will sum to 81 because each row contains the same digits, but what about the columns? Look at the first and last columns. How many times does the digit 0 appear in each column? How many times does the digit 1 appear? How many times ... Go through all 10 digits. Thinking back to the jam jar idea, this must be true for all the columns, although not as obvious as the first and last because of the muddled order of the digits.

I guess that the sum of the diagonals being 81 is just a matter of chance – or is it?

5

Two-Dimensional Shapes

For this chapter you will need:

A photocopy or tracing of figure 1
A pen, pencil and ruler
Small pieces of tracing paper, if you have them

This chapter introduces you to co-ordinates. These are pairs of numbers which enable you to locate a point on a piece of paper. There are various kinds of co-ordinates, but the most common, which are used here, will probably be familiar to you if you have ever found where a street or place is on a map.

Having drawn shapes, their symmetry is then investigated. I can remember being told at school, many years ago, that I could not – sorry: was not allowed to – use symmetry to solve problems in geometry. Feeling for symmetry seems to be natural to almost everyone, but for years, schools stuck to 'good old geometry'.

To help you draw shapes which are the same as everyone else's, you will have to use co-ordinates to locate the corners of the shapes. There are various systems of co-ordinates, but the easiest to use for this study are Cartesian co-ordinates, devised by René Descartes in the early 1600s.

In figure 1 you will see a grid of lines. Those going across the page are the x numbers, and those going up the page are the y numbers. The corner of the triangle labelled A is at the point (1,1); that is, 1 square across and 1 square up. The numbers are co-ordinated to give this point, and are therefore called the co-ordinates of the point A. Here it does not matter if you get the numbers in the wrong order because they are the same, but for corner B the numbers are different. B is at (5,1), 5 across and 1 up.

The x co-ordinate is separated from the y co-ordinate by a comma. A dot could be confused with a decimal point and nothing at all would be even more confusing. Try plotting (132535)!

The two co-ordinates are bundled together in brackets to avoid confusion with the co-ordinates of other points.

It is important to put the co-ordinates in the correct order: x first (across) and then y (up).

1. What are the co-ordinates of the point C in figure 1?

This question is for you to check your knowledge of co-ordinates before you go on.

The answer is that C is at (3,4)

Copy or photocopy figure 1 then draw these triangles on the grid:

2. Put a dot at the point (1,5), then at (5,5), then at (1,9). Join the three points with straight lines to form a triangle. Label this as triangle 2.

3. Plot the points (1,10) (3,13) (1,16) then join them to form triangle 3.

4. Plot the points $(3,15\frac{1}{2})$ (5,19) (1,19) then join them to form triangle 4.

All four triangles have a line of symmetry. For the first triangle, the one drawn for you, it is the line through C going straight down the page. Draw it, then you will see that the right-hand side of this triangle is a reflection of the left-hand side.

If you have a small piece of tracing paper, you can trace this triangle. If you then fold the paper along the line of symmetry, you will see that the left-hand and right-hand sides coincide.

5. Draw the lines of symmetry for the other three triangles, using tracing paper to locate them if you wish.

Did you puzzle over triangle 4? It is an attempt at an equilateral triangle – one with all its sides the same length. It is an attempt because it is impossible to draw an equilateral triangle exactly using Cartesian co-ordinates.

6. Look for a total of three lines of symmetry for triangle 4.

Triangle 4 also has symmetry which is not defined by a line, or lines. It has rotational symmetry of order 3. This means that you can turn the shape three times, each time through 120°, and it will occupy the same place.

You can turn any two-dimensional shape through a complete turn in its own plane and it will occupy the same place it did before you turned it. This would be rotational symmetry of order 1, except that we do not regard this as rotational symmetry.

Here are the co-ordinates of seven more shapes. They are all quadrilaterals, shapes with four sides, and it is possible – just – to draw them all on the same sheet as the triangles. Plot the points first using their co-ordinates, then join the points to form the shapes. Label each shape with the number of its question (label the parallelogram as '7') then investigate the symmetry of each one. To do this, write down, and

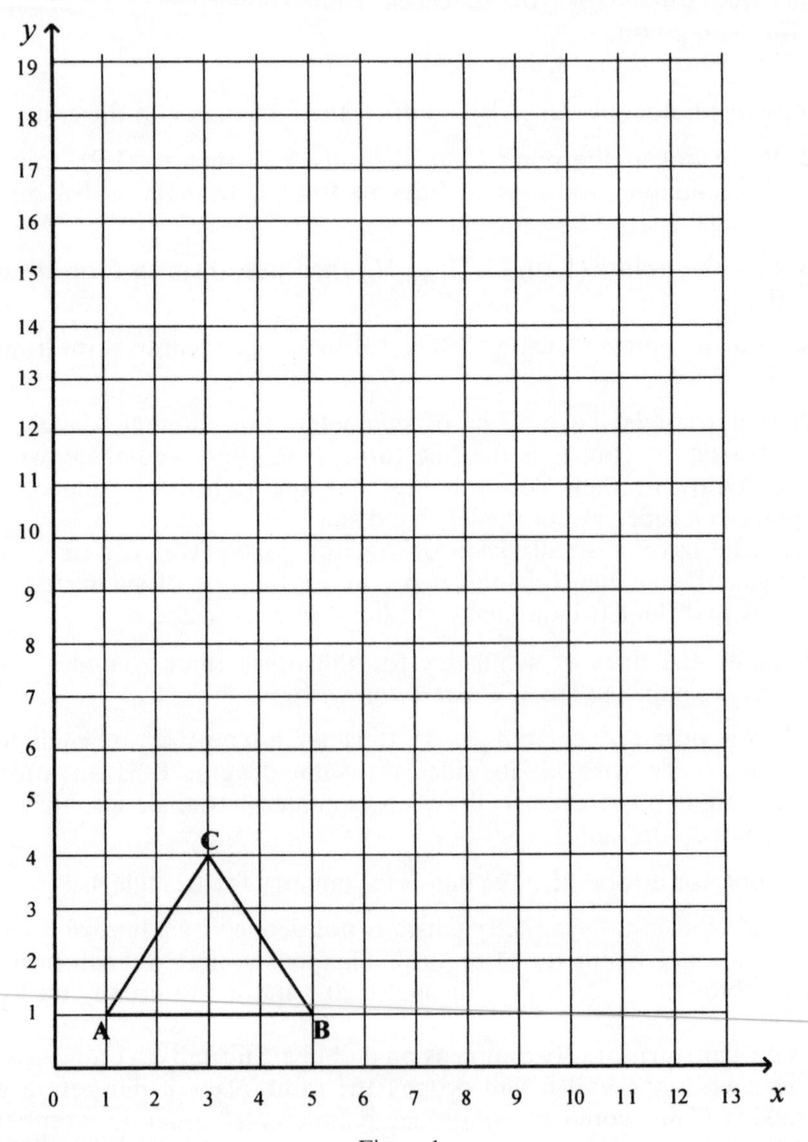

Figure 1

draw if you can, how many lines of symmetry each shape has, and also the order of rotational symmetry of each shape.

Note that some shapes may have no lines of symmetry and some may have no rotational symmetry.

7. Parallelogram: (6,1) (11,1) (13,4) (8,4)

8. Trapezium: (6,5) (11,5) (9,7) (7,7)

9. Isosceles trapezium: (4,8) (9,8) (8,10) (5,10)

10. Rhombus (or diamond): (11,8) (13,11) (11,14) (9,11)

11. Kite: (6,11) (8,14) (6,16) (4,14)

12. Rectangle: (10,15) (12,15) (12,19) (10,19)

13. Square: (6,17) (8,17) (8,19) (6,19)

Tessellations

Looking at the symmetry of shapes is one way of getting to know something about them. Another way is to tessellate them. This is mathematical jargon for 'seeing how they fit together'; literally it means 'tiling'.

Figure 2 shows a triangle with its angles labelled *a*, *b* and *c*, which has been copied to make six identical triangles (tiles) which then have been arranged so they fit about a point.

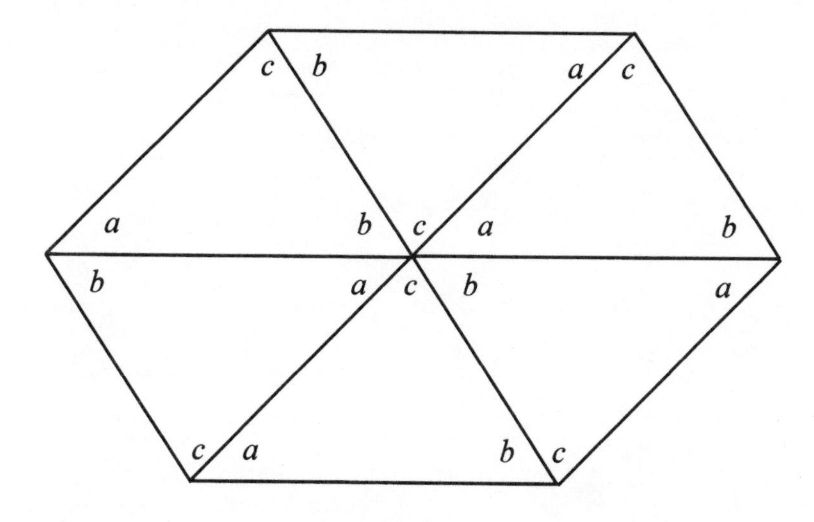

Figure 2

Below is a list of simple geometrical facts which can be found, and justified, by looking at this diagram. See if you can find them, too, and see if I have missed any.

The sum of the three angles of a triangle is 180°.
The lengths of opposite sides of a parallelogram are equal.
Opposite angles of a parallelogram are equal.
Vertically opposite angles are equal.
The sum of the interior angles of a quadrilateral is 360°.
The sum of the interior angles of a hexagon is 720°.
When a line (called a transversal) cuts two parallel lines then:
Alternate (or 'Z') angles are equal.
Corresponding angles are equal.
Interior (or 'Allied') angles add up to 180°.

We say that a shape tessellates if many copies of that shape fit together without any gaps; the edge of the compound shape may be jagged.

Any triangle will tessellate, and if you have some squared paper and an hour or so to spare, you could satisfy yourself that all the quadrilaterals you drew recently will also tessellate.

14. Will any quadrilateral tessellate? That is, if you cut out any four-sided shape and make lots of copies of it, will they fit together without any gaps? Here are two thoughts which might help: What is the sum of all the angles which fit around a point? What is the sum of the interior angles of a quadrilateral (which is two triangles)?

6

Three-Dimensional Shapes

For this chapter you will need:

Scissors and glue
Photocopies or tracings of figures 1, 2, 3, 4, and 5
An extra copy of figure 4

There is only one way to study three-dimensional shapes in my opinion and that is to make them.

As with two-dimensional shapes, it is possible to locate the corners of solid shapes by co-ordinates. You would need three co-ordinates if you drew the shapes, of course, and it is because there are three measures (length, breadth and height) that the shapes are called 'three-dimensional'.

Artists, photographers and computer graphics experts can represent three-dimensional shapes on a flat surface with great skill, but there is no substitute for handling them and being able to walk around them.

The shapes examined in this chapter are the five Platonic solids. There are relationships between various pairs of shapes and it is in examining these that you should become more familiar with the shapes themselves.

Figures 1–5 at the end of this chapter show the nets of five regular solids. A net in mathematics is the flat shape which you fold and glue to make a 3-D shape. You may need to finish off the shape with sticky tape, because it is impossible to get inside to glue the last bit.

Copy, cut out, then stick together the first two shapes, figures 1 and 2, the octahedron and the cube which is missing one face. The reason for the missing face should become clear soon.

Fold along each of the lines; fold very carefully or the accuracy of the drawings will be lost. All the folds are in the same direction; fold the paper either away from you or towards you, but be consistent.

How to examine the symmetry of a three-dimensional shape, an octahedron:

1. When looking for rotational symmetry, do you turn it about a point or about an axis?
2. Can you turn it about more than one axis to achieve rotational symmetry?
3. When looking for reflective symmetry, do you look for reflection in a line or in a plane?
4. Is there more than one plane in which the shape can be reflected symmetrically?
5. There is also something called 'point symmetry'. If you can imagine a point exactly in the middle of the octahedron, every point of the octahedron can be reflected through this point to another point of the octahedron. Look at the shape and check that this is so.
6. How is point symmetry different from plane symmetry?
7. Examine the symmetries of the cube, imagining that the sixth face of the cube is in place; in other words, do questions 1 to 5 with a cube.
8. How many faces (flat surfaces) has the octahedron?
9. How many corners has the octahedron?
10. How many edges has the octahedron?
11. Add the number of faces to the number of corners then take away the number of edges. (Do: $F + C - E$.)
12. How many faces has the cube (if it were complete!)?
13. How many corners has the cube?
14. How many edges has the cube?
15. Add the number of faces to the number of corners then take away the number of edges. (Do: $F + C - E$ again.)

Questions 11 and 15 illustrate Euler's Relation: Faces + Corners – Edges = 2. Euler lived and worked in the eighteenth century.

You may also notice that the number of faces and the number of corners of the two shapes swap over. The number of edges is the same for both. This property defines the two shapes as 'duals'.

You will now find out why the cube is missing a face. You should be able to fit the octahedron inside the cube so that each of its corners touches the middle of each face of the cube. This shows that the number of corners of an octahedron is the same as the number of faces of a cube.

You may also realise that you could fit a cube inside the octahedron so that each corner of the cube touched the middle of each face of the octahedron. I thought this was too difficult to do with small pieces of paper.

16. Copy, cut out and glue together the icosahedron and dodecahedron. The icosahedron has 20 faces (*eikosi* is Greek for 'twenty') and is straightforward to make, but rather fiddly at this size.

The net of a dodecahedron (12 faces; *dodeka* is Greek for 'twelve') is very difficult to fit on to a page without being extremely small, so you will have to build it up slowly, and in two halves. You will need two copies of figure 4, each one making half the dodecahedron. Stick each of the sides to one edge of the base, then stick the sides to each other; each half should look something like a fruit bowl. Then stick both halves together.

17. How many faces has a dodecahedron?

18. How many faces has an icosahedron?

The dodecahedron and the icosahedron are duals – the number of their faces and corners swap.

19. How many corners has a dodecahedron?

20. How many corners has an icosahedron?

Euler's Relation says that the number of faces + the number of corners – the number of edges = 2.

21. Use Euler's Relation to find the number of edges of a dodecahedron.

22. How many edges has an icosahedron?

It is probably easier to calculate these answers than it is to count them! You can count them to check if you have enough patience.

23. Now copy, cut out, then glue together the tetrahedron, figure 5. The net is in two pieces so that the shape can be large and its net still fitted on to one page. One tab is missing because the last one is always impossible to glue from the inside. The final shape should be a triangular-based pyramid.

24. Count the number of faces, corners and edges and check Euler's Relation.

25. What shape is the dual of the tetrahedron?

The five shapes you have made up are called the Platonic Solids, although I am not sure what they have to do with Plato. They are solids which can be made up from lots of regular polygons. (A polygon is a flat shape with several straight line sides; it is said to be regular if all these sides, and the angles between them, are all the same.)

26. In a plane, that is, on a flat surface, you can always draw a circle which touches all the corners of any triangle; it is called the circumcircle. Can you make a sphere which touches all the corners of any of the Platonic Solids?

27. See if you can think of a simple way to check your answers to question 26. (Hint: a sphere will always have a constant diameter.)

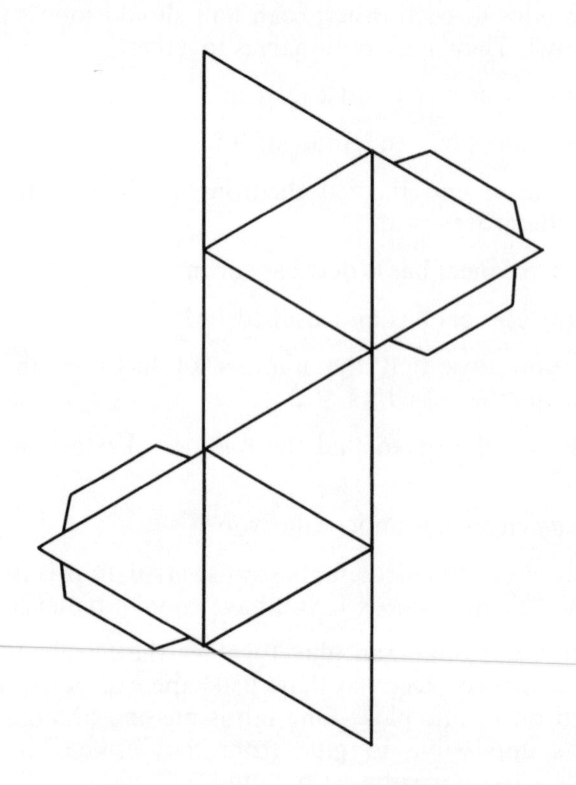

Figure 1: The net of an octahedron

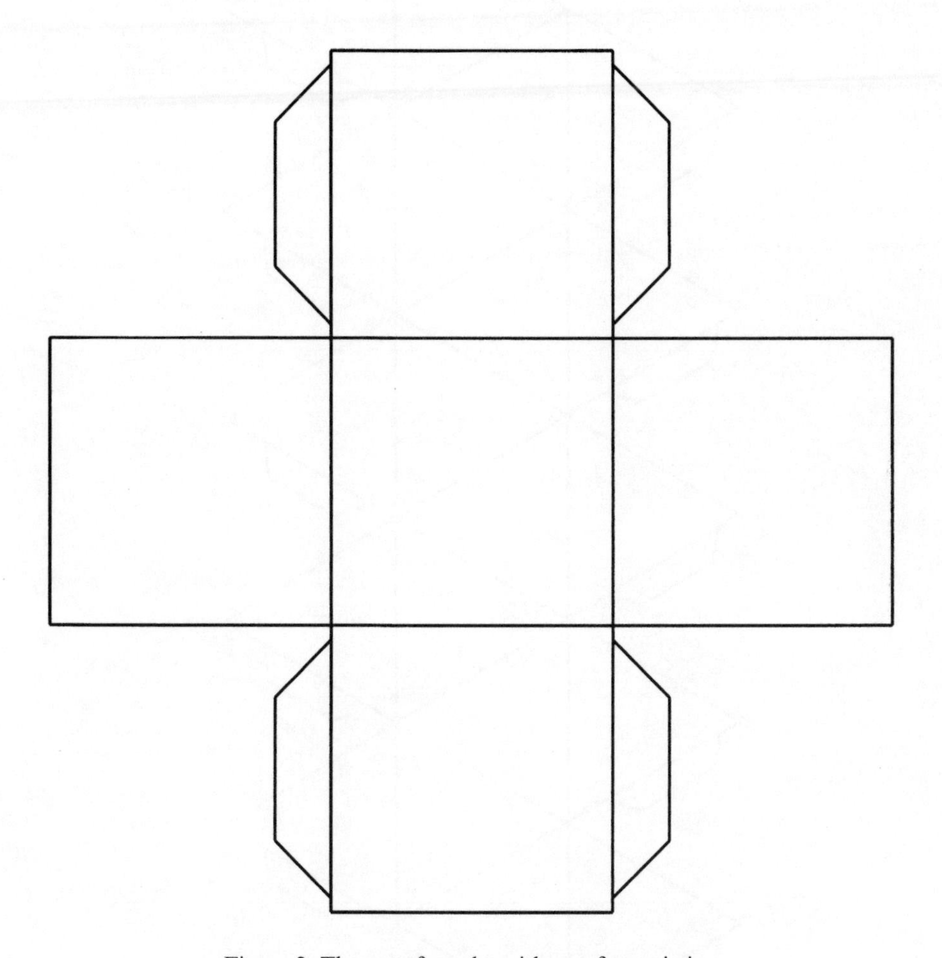

Figure 2: The net of a cube with one face missing

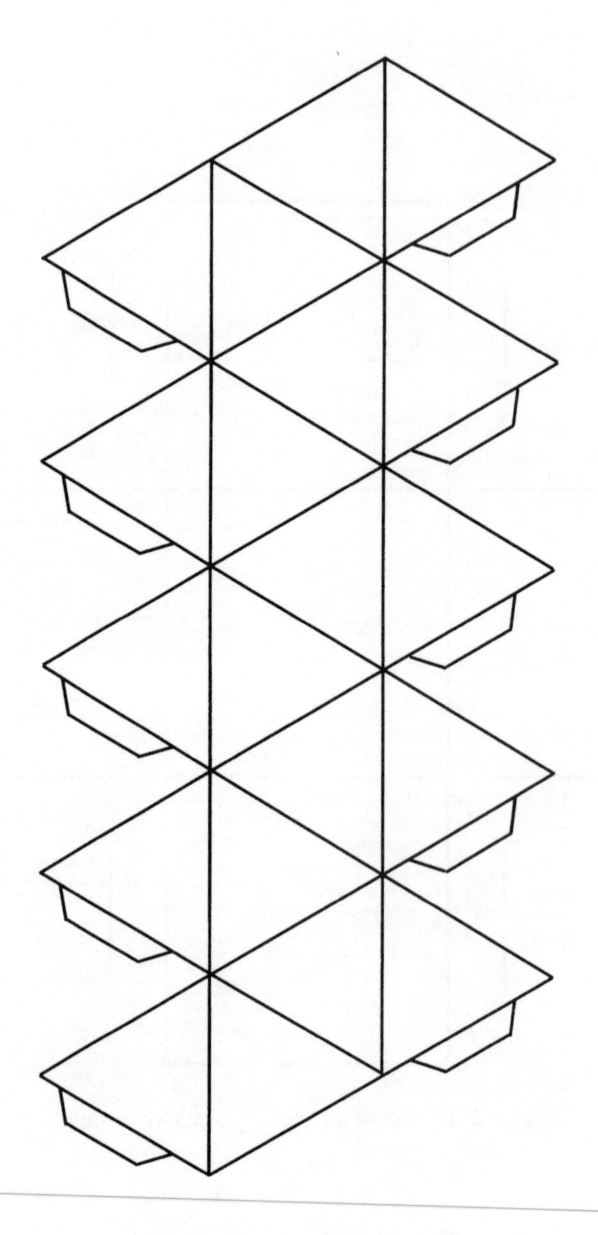

Figure 3: The net of an icosahedron

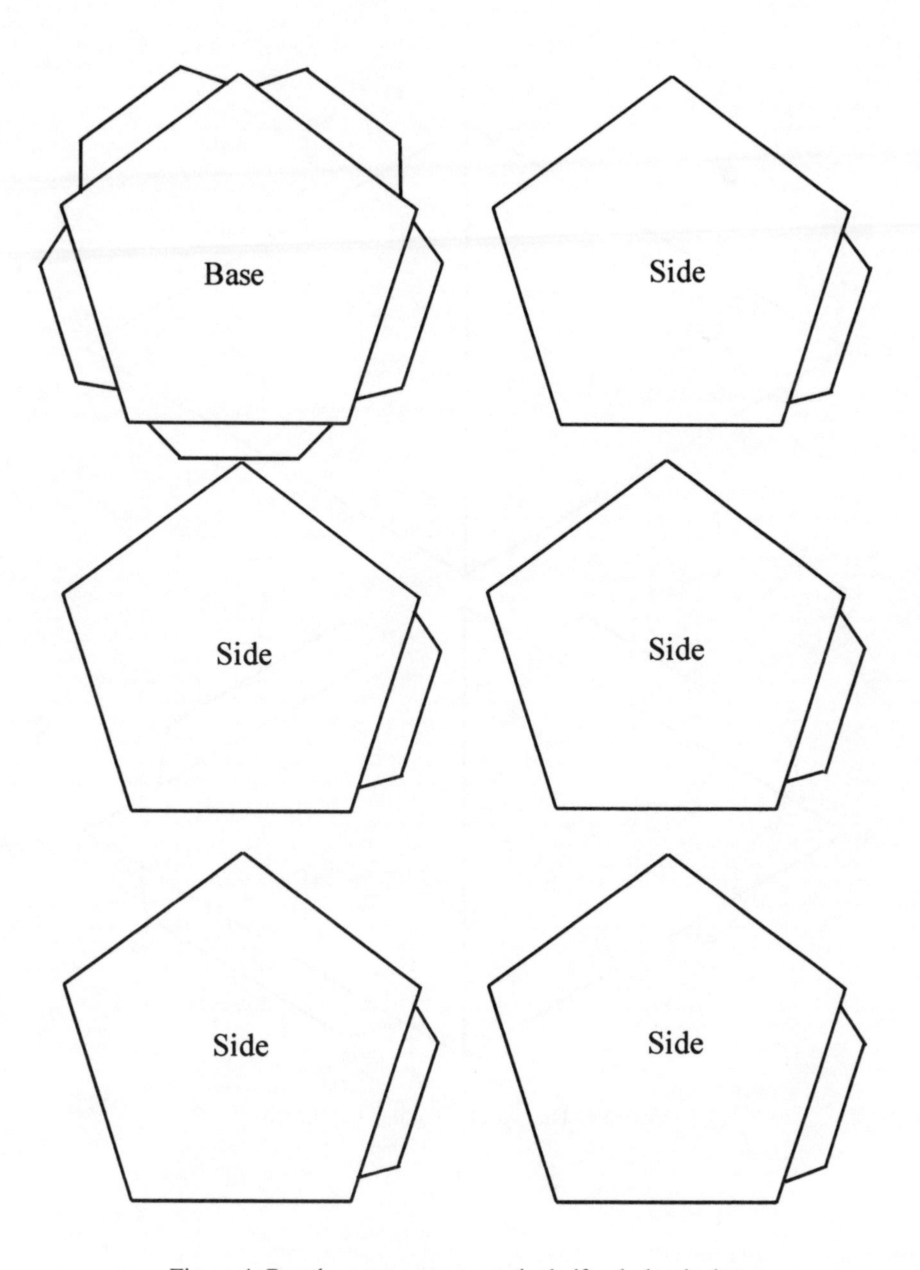

Figure 4: Regular pentagons to make half a dodecahedron

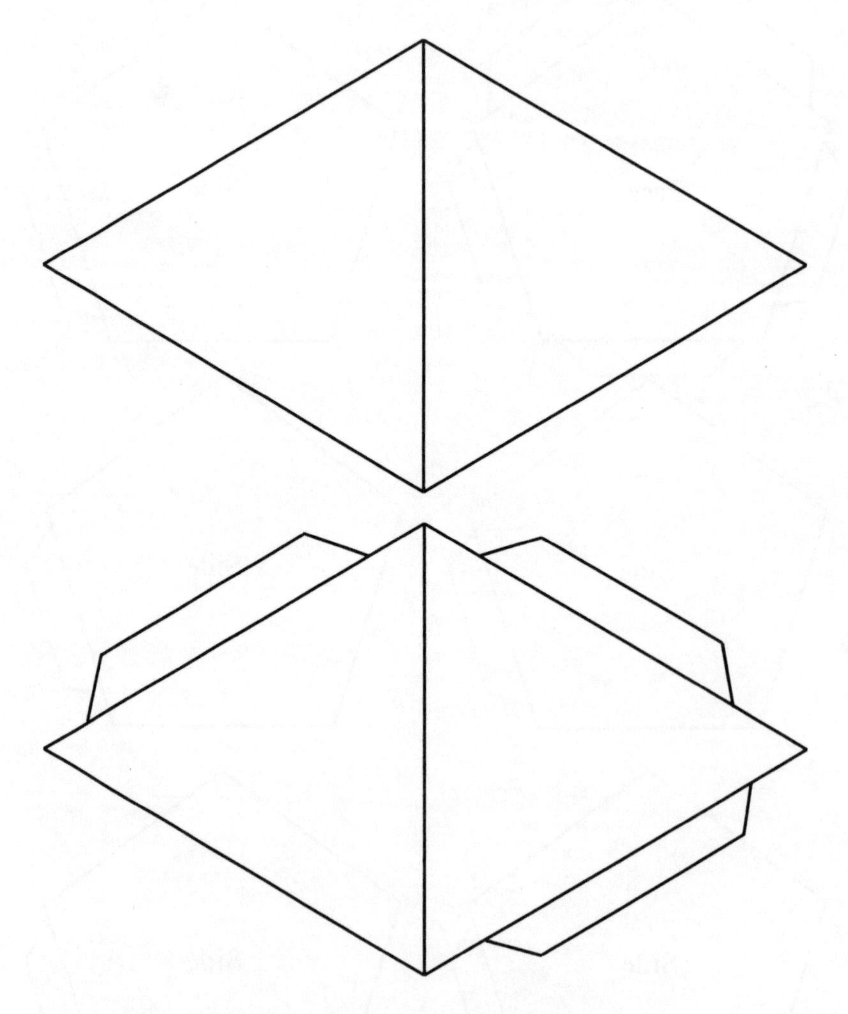

Figure 5: The net of a regular tetrahedron

7

Pascal's Triangle

For this chapter you will need:

Pen and paper
A calculator if you wish to use one

Pascal's Triangle is fairly easy to construct, but as with so many other simple structures in mathematics, it has many patterns in the numbers and also some useful applications.

Some of the examples given are rather contrived; did you ever hear of 24 entrants to an athletics event who all ran 100m in exactly 12 seconds? Your suspension of disbelief has to be extended further when the meeting starts, because, of course, there are no dead heats!

However, I hope you can understand the reason for including heats and finals of an athletics meeting: in the heats, the first three go through (Why is it three? There is a good reason.) regardless of the order in which they finish, whereas in the final the order of the first three particularly is all-important.

When you do questions 11 to 21, you are not asked to do them experimentally, although that is probably the best way. If the question asks: 'In how many ways can you select two things from five', put five things in front of yourself and experiment to find the answer. The world is a real place; there is nothing wrong with making it real when it helps to solve mathematics problems.

I have heard a story that many years ago, a French king, probably called Louis, loved gambling, but hated losing. He asked the mathematician Blaise Pascal to investigate why he lost so much. In the course of his work, Pascal came up with this triangle of numbers, which was known in China some centuries before.

The triangle has been started for you with lots of guidelines to help you see how it is formed. Each line begins and ends with a 1, and each number comes from adding the two numbers nearest to it in the row above.

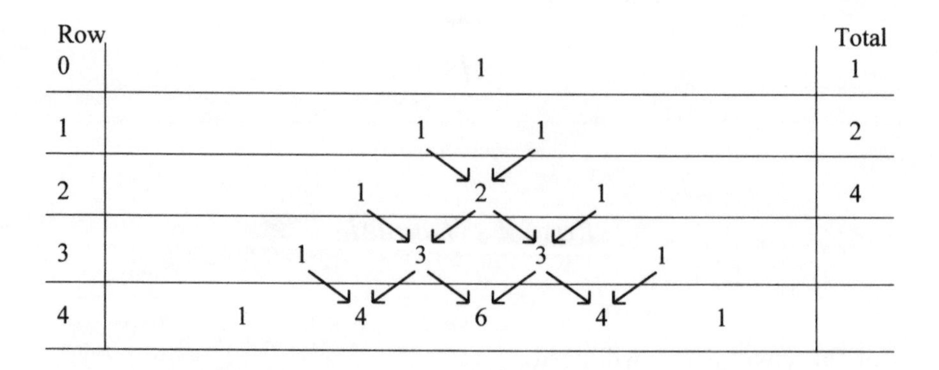

Figure 1

1. Make your own copy of Pascal's Triangle and continue it until you run out of room.
2. Number each row, starting as a computer would, at 0.
3. Add up the numbers in each row, writing your answers on the right.
4. Try to find out how the total of the numbers in each row is connected to the number of the row, i.e. how is 8, the total of the numbers in row 3, connected to 3?
5. Here is a piece of mathematical shorthand: $2 \times 2 \times 2$ can be written as 2^3. Note that this is *not* the same as 2×3. So, how do you write $2 \times 2 \times 2 \times 2 \times 2$ in shorthand?
6. Look to see why the total of the numbers in each row doubles as you go down; it is obvious when you see it!
7. The first diagonal you see – sloping to the left or to the right – consists of 1s. The next diagonal consists of the counting numbers – or natural numbers. Can you see how the numbers in the next diagonal follow each other in a pattern? They are: 1, 3, 6, 10.

This is all very interesting, but we must now make use of Pascal's Triangle.

Hold up one hand with fingers and thumb spread out. For the next page or so I will call the thumb a finger – the language is easier that way. I am assuming that you have four fingers and a thumb; if not, please imagine that you have.

8. With the index finger of your other hand, touch a fingertip of your spread-out hand. How many different fingers can you touch?

9. You may have to work on this question for a while. Suppose you now touch two fingertips. In how many ways can you do this? Hint: Letter your thumb A and your fingers as B, C, D and E. Touch – and count – two in this order: A & B; A & C; A & D; A & E; B & C; B & D; etc. etc...

10. Here is a question which may seem pointless at first: in how many ways can you touch no fingers?

Here are questions 10, 8 and 9 – in that order – asked again using slightly different language: 'touching' now becomes 'choosing' a finger.

11. If you have 5 things to choose from, in how many ways can you choose none?

12. If you have 5 things to choose from, in how many ways can you choose 1?

13. If you have 5 things to choose from, in how many ways can you choose 2?

Now look at row 5 of Pascal's Triangle. Its first 3 numbers should give your answers to questions 11, 12 and 13.

14. Using a few ditto marks, what would be the next three questions in the sequence started by questions 11, 12 and 13 ?

15. Look at row 5 of Pascal's Triangle to find the answers to the three questions you have just made up.

Here is 'the light' in case you have not seen it yet! Every time you choose 2 things from 5, you also choose 3 things – the 3 things left behind. Try it with your fingers again: touch two fingertips and you leave three fingertips untouched. So, the number of ways of choosing 2 things from 5 is the same as the number of ways of choosing 3 things.

16. If you touch 1 finger out of 5, how many fingers are untouched?

17. If you do not touch any fingers out of 5, how many fingers are left untouched?

Here are a few questions to practise the use of Pascal's Triangle:

18. If you pull a coin from your pocket and place it flat on a table (a) how many different sides can face up? (b) how many of these faces are 'heads'? (c) how many of these faces are 'tails' (i.e. not heads)?

19. If you place two coins on a table, in how many ways can they be (a) both heads? (b) both tails? (c) a head and a tail? (This means H with the first coin, T with the second, then – differently – T with the first coin and H with the second.)

20. Suppose you place three coins on a table; write down all the possible arrangements of heads and tails (e.g. as HHH or HTH etc. noting that HTH is not the same as HHT because a different coin is tails up). Use row 3 of Pascal's Triangle to make sure you have all the possibilities.

21. Seven tennis players arrive at a tennis club one day and agree to play doubles. In how many ways can they pair off in twos?

Joe Leggit has entered the 100m in the Tricketts Cross Athletics Event. He is one of 24 entrants, who all have a personal best for the season of exactly 12 seconds.

'It will all come down to luck, and who is best on the day,' said Joe.

There are 4 heats in round 1, each with 6 sprinters; the first 3 over the finish line go through to the next round.

22. Does it matter in which order the first three finish in round 1?

23. How many different groups of 3 can go through from heat 1, round 1?

24. How many sprinters will go into the semi-final?

25. What fraction of the original entrants go into the semi-final?

26. The semi-final follows the same rules as the previous round, leaving 6 sprinters in the final. Does it matter in which order they finish in the final?

Note that the order in which the first three finish in the final is important – at least to the runners – whereas the order in which the first three finish in the heats is not important; all that matters to a runner is that he is in the first three.

Pascal's Triangle gives the number of ways in which three runners can be selected from six taking no account of the order in which they are chosen. Consequently, you could use it to find how many ways the heats could finish, but not the number of ways in which the final could finish.

Note, too, the reason for the stupidity of having 24 runners who all run at the same speed. It means that at the beginning of a race any one of the six starters could be the winner. In mathematics we say that all the various results of each race are 'equally likely' – a condition which is necessary when you apply Pascal's Triangle. It is obvious that if Joe could run 100m consistently in under 11 seconds when everyone else ran in 12, he would always win; the results of his races would not be equally likely.

8

Probability

For this chapter you will need:

Pen and paper
A calculator

After mastering the skill of calculating the number of ways of selecting things and of doing things, we move on to what is really part of statistics.

Calculations in probability assume that we live in a perfect world, a world where a coin will come down heads half the times it is tossed, and, of course, tails for the other half. It is a reasonable assumption that it will not come down and rest on its rim.

The next stage of statistics is to gather and display information. That side of things is only touched upon in the next chapter. Here all I need to say is that you have to know what you are trying to determine before you ask the questions. The questions then have to be suitable for analysis; for example, asking 'What is your favourite breakfast?' is not a very helpful question because you are likely to get hundreds of different answers, and what can you make of them? If you are doing a survey for a cornflakes company, then the question would obviously be: 'Do you have cornflakes for breakfast?'

Having gathered a mountain of useful information, a statistician then has to analyse it, and part of that analysis is to see how the information agrees with what you might expect in a perfect world. That is where the perfect world of probability comes in.

I have used a tree diagram for some of the work. You have to turn it on one side before it looks anything like a tree. But a diagram sorts out information better than anything else I know. We also re-enter the world where all entrants in a race have the same personal best time, but there are no dead heats!

As with all the other chapters, you may find the work becoming difficult towards the end, but as I am constantly reminding you, this is not a course with an examination at the end of it, so if you cannot manage to solve a problem it does not matter. The only thing which suffers is your pride, but since I think you deserve a medal for trying, take pride in that!

WARNING: Some of my original group of students found this chapter to be very difficult, particularly towards the end. Don't continue with it if you do.

1. If 6 runners, including Joe Leggit, of equal ability, start a race in round 1, and 3 of them go through to round 2, what is the probability that Joe goes into round 2?

Assuming that race positions are equally likely – a piece of mathematical fiction, of course – there is only one answer to this question, but several ways of writing it. 3 out of 6 runners go through to round 2, so the probability that Joe is one of them is

3 out of 6, or $\frac{3}{6}$ or $\frac{1}{2}$ or 0.5 or 50%

As you will realise later, it is not a good idea to answer 'evens', and 1–1 is wrong. Horse race odds of, say, 10–1 mean that a horse is likely to lose 10 races to every 1 it wins, giving a probability of winning as $\frac{1}{11}$.

Here is some practice:

2. Imagine you draw a card from a shuffled pack. (a) How many aces are there in the pack? (b) How many cards are there in the pack? (c) What is the probability that it is an ace?

3. What is the probability that a card drawn from a pack is a diamond?

4. What is the probability that the card is the ace of diamonds?

5. What is the probability that when you roll a fair die (plural 'dice') you get a 6?

6. In question 2, there are 4 aces in a pack of 52 cards. The probability of drawing an ace is $\frac{4}{52}$ or $\frac{1}{13}$. What is the probability of drawing a card which is not an ace?

7. In question 3, there are 13 diamonds in a pack of 52 cards. The probability of drawing a diamond is $\frac{13}{52}$ or $\frac{1}{4}$. (Note that $\frac{1}{4}$ of the pack are diamonds.) What is the probability of drawing a card which is not a diamond?

8. In question 4, there is 1 ace of diamonds in a pack of 52, so the probability of drawing the ace of diamonds is $\frac{1}{52} = \frac{1}{13} \times \frac{1}{4}$, the answers to questions 2 and 3. What is the probability of drawing a card which is not the ace of diamonds?

9. In question 5, there are 6 faces on a die, 1 of which is a 6. The probability of rolling a 6 is $\frac{1}{6}$. What is the probability of a score which is not 6?

Here are two questions which may seem silly:

10. If you roll a normal, fair die, what is the probability that you score a number between 1 and 6 inclusive?

11. If you roll the same die, what is the probability that its score is 9?

The purpose of questions 10 and 11 is to show that the probability of certainty is 1 or 100%, and the probability of impossibility is 0.

Note also, looking at questions 6, 7, 8 and 9, that if the probability of something happening is, say, $\frac{1}{4}$, the probability of it not happening is $\frac{3}{4}$ or $1 - \frac{1}{4}$.

The fun with probability starts when probabilities are combined.

12. If a card is drawn from a shuffled pack, what is the probability that it is an ace?

13. If another card is then drawn from this pack, what is the probability that it, too, is an ace?

Question 13 cannot be answered until a couple more questions have been answered.

14. If the first card is replaced in the pack and the pack shuffled, (a) how many aces are in the pack? (b) how many cards are in the pack? (c) what is the probability that the second card is an ace?

15. If the first card drawn is an ace, and it is *not* put back in the pack, (a) how many aces are in the pack? (b) how many cards are in the pack? (c) what is the probability that the second card is an ace?

16. If the first card drawn is not an ace, and it is *not* put back in the pack, (a) how many aces are in the pack? (b) how many cards are in the pack? (c) what is the probability that the second card is an ace?

If the first card is put back into the pack, it does not matter what happened when the first card was drawn, the probability of an ace remains the same. Here we say that the two probabilities – that the first card is an ace and that the second card is an ace – are *independent*.

In questions 15 and 16 the probability that the second card is an ace depends on what happened when the first card was drawn. Here, just to be awkward, we do not say that the probabilities are dependent, but that we have conditional probability – the second probability is conditional on what happened before.

Life becomes complicated when we draw tree diagrams to show conditional probabilities. I will keep to something more straightforward, so here is a tree diagram to show all the possibilities when you try to get two consecutive 6s when rolling a die.

Here are some questions by way of reminders:

17. What is the probability of getting a 6 when you roll a die?

18. What is the probability of getting a number which is not a 6 when you roll a die?

Rolls of a die are independent events, so the second roll of a die does not depend on what happened the first time.

19. If you get a 6 with the first roll of a die, what is the probability that you get a 6 with your next roll?

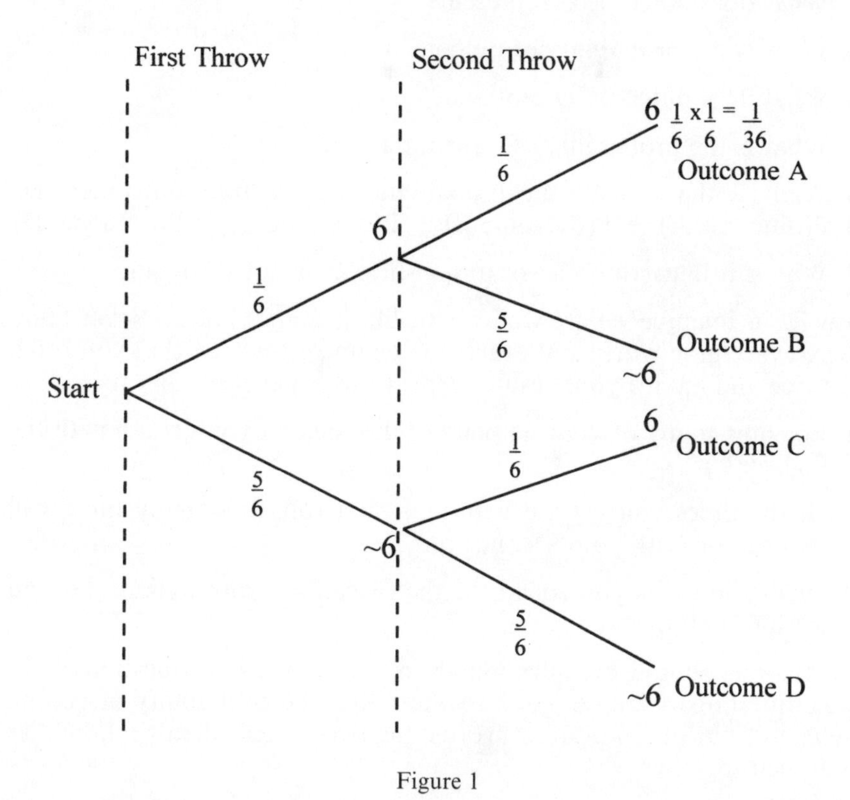

Figure 1

Figure 1 is the tree diagram which shows the four possible outcomes of rolling a die twice when hoping for a 6 both times.

The symbol ~6 means 'not a 6'.

Outcome A is at the end of two branches of the tree and is what we hoped for: two sixes, one after the other. The probability of each event is written beside its corresponding branch, and so the probability of outcome A is: $\frac{1}{6} \times \frac{1}{6} = \frac{1}{36}$

The reason why we multiply the two probabilities will be justified soon.

20. What does outcome B represent?

21. What is the probability of outcome B ?

Answers: check before you go on. Question 20: outcome B represents a 6 with the first roll of the die, and not a 6 with the second. Question 21: the probability of this is: $\frac{1}{6} \times \frac{5}{6} = \frac{5}{36}$

22. What does outcome C represent?

23. What is the probability of outcome C?

24. What does outcome D represent?

25. What is the probability of outcome D?

26. What is the sum of the probabilities of the four outcomes? i.e. P(outcome A) + P(outcome B) + P(outcome C) + P(outcome D)

27. Why is it that your answer to question 26 is not surprising?

Now let us imagine we are working in Mathematics Never-Never Land where everything works out exactly as planned! Imagine that you roll a die twice and record your results; repeat this experiment 36 times.

28. For how many of your 36 pairs of die-rolls do you get a 6 with the first die?

29. Of the times you get a 6 with your first roll, how many times will you get a 6 with your second roll?

30. Of the 36 times you rolled the die twice, how many times did you get a 6 both times?

The answers should be: question 28: 6; question 29: 1; question 30: 1. These questions were designed to show that the probability of getting two 6s is 1 out of 36, in other words, the two independent probabilities multiplied together.

Here are some romantic questions which combine Pascal's Triangle with probability.

Six young executives go on an adventure training weekend. One of them is Charles, who is very, very shy and desperately keen to be paired off for the overland hike with Fiona. Unbeknown to Charles, Fiona, who is also very, very shy, is desperately keen to be paired off with Charles.

31. In how many ways can the six young executives be paired off?

32. In how many of these pairings will Charles and Fiona be together?

33. What is the probability that Charles and Fiona will both have their dreams come true?

None of the questions has really been about gambling despite the introduction to the work on Pascal's Triangle. This is because I am too mean to gamble and so know nothing about it.

9

Statistics

For this chapter you will need:

Pen and paper
A friend or two
Photocopies of figures 1 and 2 (Figure 2 is not easy to trace.)

Although you may be having doubts these days about Father Christmas, I am sure you have no doubts that Mathematical Never-Never Land does not exist. Whatever results Probability Theory may predict, they rarely come true. The purpose of statistics is to gather data and see how well it fits predictions made beforehand.

The work of this chapter touches on the fact that the more data you collect, the more reliable it is – as a general rule, at least. Data you collect is then used rather than simply made into a pretty chart and displayed on your wall!

The letter frequency data is used to decode messages. It was the need to decode messages during the Second World War which led to the invention of the electronic computer. In the text, and in the code, I have given credit to Alan Turing on whose ideas the computer was based. Much of its construction was down to Dr Tommy Flowers, an electronics engineer from the Post Office. My decoding process is much simpler than the one used at Bletchley Park during the war, but at least you do not have to design and build a computer to solve the puzzle!

1. Take a short paragraph (in English) and count how many times the letter 'a' is used, how many times the letter 'b' is used, how many times the letter 'c' is used, and so on through the alphabet.

Below is the result of a similar letter frequency count, but on a longer passage; I took 20 lines of a book.

Letter:	a	b	c	d	e	f	g	h	i	j	k	l	m
Number:	69	10	24	27	99	22	16	40	46	3	4	41	15
Letter:	n	o	p	q	r	s	t	u	v	w	x	y	z
Number:	62	68	27	0	49	47	81	16	7	18	0	14	1

2. Draw the information from the last table as a bar chart on a copy of the axes of figure 1. You will have to fill in the numbers on the vertical axis, and you will have to approximate the height of most of the bars.

3. Draw your own results from question 1 on the same graph. If you use a colour different from the one you have already used it will distinguish one graph from the other.

Compare your results with the ones in my table. Here is a simple way to compare: write down your 'top 5' in order of popularity. The top 5 from the results above are: e, t, a, o, n. The graph will almost certainly show that our two sets of results are different.

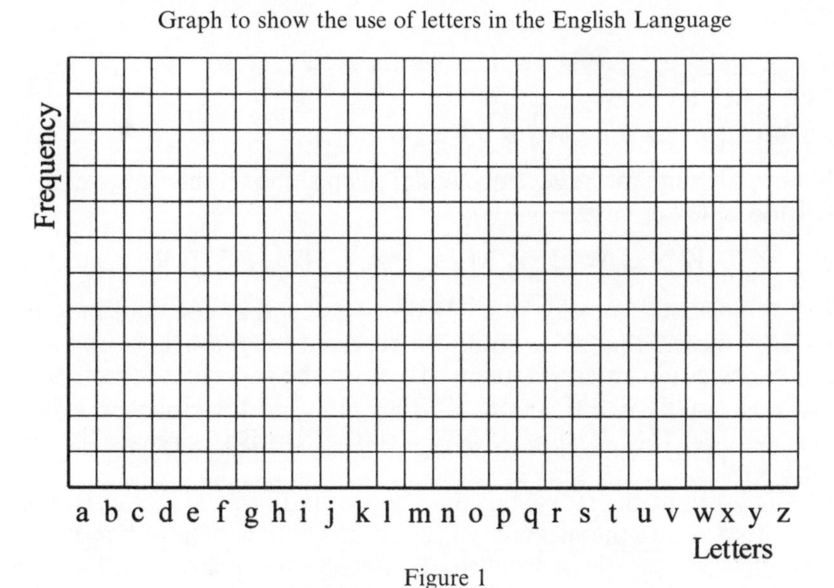

Figure 1

It is likely (although not certain) that the results of my table are more representative of the 'complete' English language, not because I am cleverer than you – I am almost certainly not! – but because I used a larger sample. If you were to find the letter frequency for hundreds of books by various authors you would get a result far more reliable than mine.

We are now going to extend these results showing letter frequency to codes.

Here is a letter substitution cipher (sometimes 'cypher') for making up a code:

A	B	C	D	E	F	G	H	I	J	K	L	M
b	d	k	p	y	l	s	x	a	g	j	r	m

N	O	P	Q	R	S	T	U	V	W	X	Y	Z
u	w	e	i	o	n	t	c	v	z	h	q	f

The alphabet is written out in the correct order in the top line, in CAPITALS, and underneath, it is written out again in a completely jumbled way, but in lower case (small letters). Later on there is a much more difficult code to decipher, so it is a good idea to stick to the rule that the true message is in capitals and the coded message in lower case. Here is a coded message which used the cipher above. Finish it!

4. q w c m c n t d y v y o q k r y v y o a l q w c k b u
YOU MUST
p y k w p y t x a n m y n n b s y.

5. Now put this message into code, using the cipher above and writing in lower case:

Y E S I K N O W I A M V E R Y C L E V E R.

During the Second World War, Hitler's army sent its messages by radio. This meant that they could be received very quickly, but that their enemies could intercept them. To keep the messages secret, they used a code exactly like this one, but they changed the cipher to a new one every day. They were known to the British services as the 'Enigma' codes.

There is more to the story, but in order to find out the rest of it, you will have to de-code the message of figure 2. There is help at hand:

The coded message is in English. To decode it quickly, you have to have an idea of the frequency with which letters are used in the English language. During the Second World War, British Army decoders would have needed to know the frequency with which letters were used in German (and much more besides). From your earlier work, you have a good idea of English letter frequency.

Next, you must know the frequency with which letters occur in the y v d / d i m f e j v / g x n l / v g q / y o / z e i q / g / k g l /

_ _ _   _ _ _ _ _ _ _   _ _ _ _   _ _ _   _ _   _ _ _ _   _   _ _ _

o z / r x d g w e i m / y v d / h o q d j , / j o / y v d l /

_ _   _ _ _ _ _ _ _ _   _ _ _   _ _ _ _ _   _ _   _ _ _ _

g j w d q / g / n g i / h g f f d q / g f g i / y b x e i m / y o /

_ _ _ _ _   _   _ _ _   _ _ _ _ _ _   _ _ _ _   _ _ _ _ _ _   _ _

q o / e y / z o x / y v d n . / v d / k o x w d q / k e y v /

_ _   _ _   _ _ _   _ _ _ _   _ _   _ _ _ _ _ _   _ _ _ _

f o y j / o z / j h e d i y e j y j / g i q /

_ _ _ _   _ _   _ _ _ _ _ _ _ _ _ _   _ _ _

n g y v d n g y e h e g i j , / g i q / e i / j e u / n o i y v j /

_ _ _ _ _ _ _ _ _ _ _ _ _ _   _ _ _   _ _   _ _ _   _ _ _ _ _ _

y v d l / v g q / n g q d / g / n g h v e i d / k v e h v /

_ _ _ _   _ _ _   _ _ _ _   _   _ _ _ _ _ _ _   _ _ _ _ _

h o b f q / r x d g w / y v d / d i d n l / h o q d . / e y /

_ _ _ _ _   _ _ _ _ _   _ _ _   _ _ _ _ _   _ _ _ _   _ _

k g j / y v d / z e x j y / h o n a b y d x / e i / y v d /

_ _ _   _ _ _   _ _ _ _ _   _ _ _ _ _ _ _ _   _ _   _ _ _

h o x f q / g i q / k g j / h g f f d q / y v d /

_ _ _ _ _   _ _ _   _ _ _   _ _ _ _ _ _   _ _ _

h o f o j j b j . / e y / y o o w / b a / n o j y / o z / g /

_ _ _ _ _ _ _ _   _ _   _ _ _ _   _ _   _ _ _ _   _ _   _

f g x m d / x o o n / r b y / k g j / i o y / i d g x f l /

_ _ _ _ _   _ _ _ _   _ _ _   _ _ _   _ _ _   _ _ _ _ _ _

g j / a o k d x z b f / g j / y v d / q d j w - y o a /

_ _   _ _ _ _ _ _ _ _   _ _   _ _ _   _ _ _ _   _ _ _

h o n a b y d x j / k d / b j d / y o q g l .

_ _ _ _ _ _ _ _ _   _ _   _ _ _   _ _ _ _ _ .

Figure 2

coded message. To help you in this – if you want the help – there is a letter-frequency list for the coded message below.

It is likely, though not certain in such a small sample, that the most frequently used letter in the code stands for the letter E in 'real English'.

You can look for other frequently used letters, but it is also a good idea to look for commonly used words, particularly the first one of the message.

Single-letter words are also something of a give-away.

Space has been left below each letter for you to write in – preferably on a photocopy – what you think the correct letter should be.

Here is a table showing the frequency with which the letters are used in the coded message, *but* it may not be 100% reliable; nothing ever is! I would suggest that you use it as a guide; for example, 'd' is the most frequently used letter in the code, so it is likely, although not certain, that it corresponds to 'E' in the decoded message.

Letter:	a	b	c	d	e	f	g	h	i	j	k	l	m
Frequency:	5	9	0	33	18	13	32	12	19	23	10	7	4

Letter:	n	o	p	q	r	s	t	u	v	w	x	y	z
Frequency:	13	28	0	18	3	0	0	1	21	6	14	32	7

Note: There is at least one mistake in the code in Figure 2, but there might well have been mistakes in the codes which they had to work on during the war.

10

Conic Sections

For this chapter you will need:

Pen and paper
Scissors and glue
Photocopies or tracings of figures 1, 2, 4, 5 and 8

Conic sections, as the name implies, are sections of a cone. This means that if you can imagine a cone sliced with a clean cut from a large knife, the surface, or the edge of the cone which is exposed by the cut is a conic section. Most of the shapes will be familiar to you.

A circle is one obvious section of a cone, and very familiar. I have been going round in circles for years! However, just to be awkward, I have not done any work on circles in this chapter, probably because they are so familiar.

The ellipse is the next one to come across, as you angle your knife just a little steeper. You may not realise it, but we have all been going round in an ellipse for years: the Earth orbits the Sun in an ellipse, although it is very close to a circle in its shape.

The parabola is the other conic section to be looked at in this chapter. As the work will reveal, there are plenty of parabolas around us although they are not always as easy to recognise as circles and ellipses.

Two other conic sections have been all but ignored: a pair of straight lines and a hyperbola. Their applications are obscure, and they are also difficult to demonstrate as sections of a cone. Indeed, the cone has to be a double one, as explained in the text. I hope you are not disappointed that they have been left out!

It's amazing what you can do by cutting up a cone. In figure 1, you will see some nets of cones which have been cut. A net in mathematics is the flat piece of paper which you fold to make a three dimensional object.

1. Trace or photocopy figures 1(a) and 1(b), cut out the shapes and stick them together.

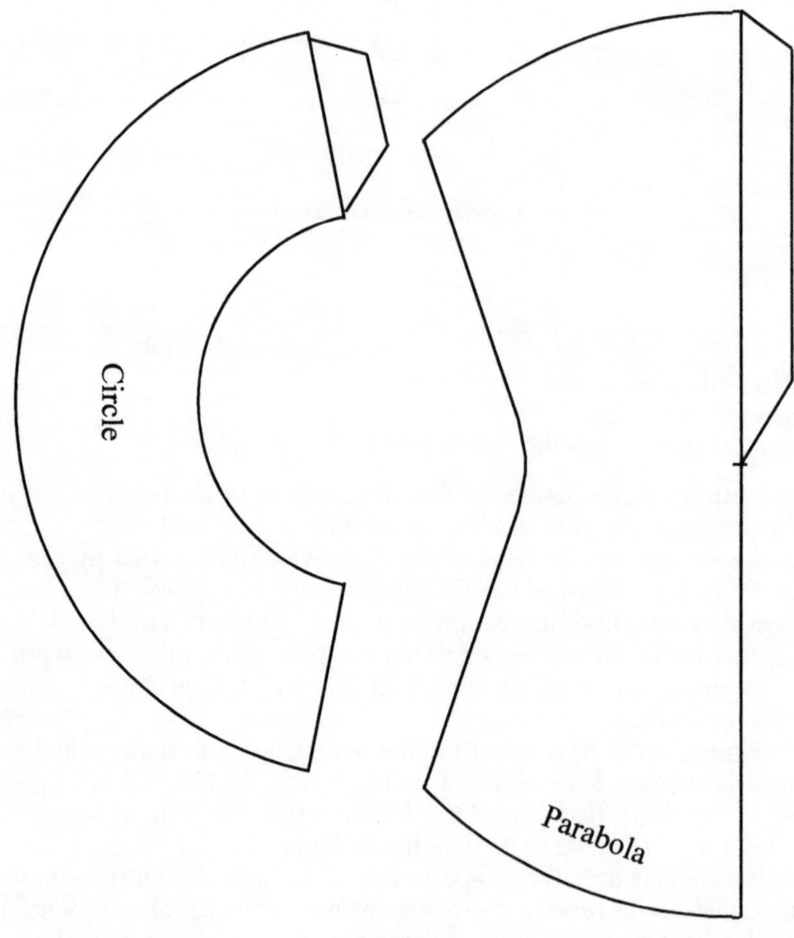

Figure 1(a)

The circle: If you cut a cone in a plane perpendicular to the cone's axis of symmetry, you get a circle. We will do some more with circles in Chapter 17.

The ellipse: The word 'ellipse' is given in my dictionaries as coming from Greek words for 'come short' and 'a defect'. Presumably it is a defective circle. You get it from a cone by sloping your cut across the cone a little.

The parabola: This is given as 'a throw beside' or 'placing side by side'. You have to cut a cone by a plane which is parallel to the sloping edge of the cone.

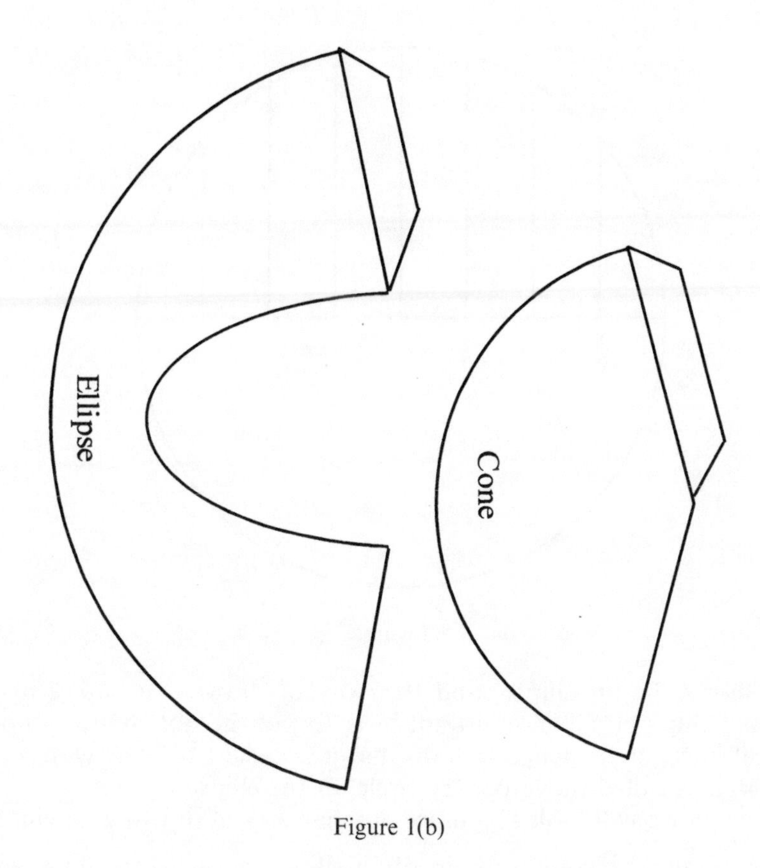

Figure 1(b)

The hyperbola: This is given as meaning 'a throw in excess' or simply 'in excess'. This probably refers to the angle of the cut, which has to be steeper than that for the parabola. Also, the hyperbola has two parts or branches; this is because the cone here is a double cone, rather like a diabolo toy, which the cut intersects in two distinct but identical parts. The hyperbola is not shown here.

The Ellipse:

2. Draw, trace or photocopy a circle and add a diameter across, and a few lines going up from this diameter to the circle as in figure 2. Measure the length AB then put a dot half way up AB. Put another dot the same distance from the diameter, but directly below A. Do the same for all the other vertical lines, then join the dots with the smoothest possible curve.

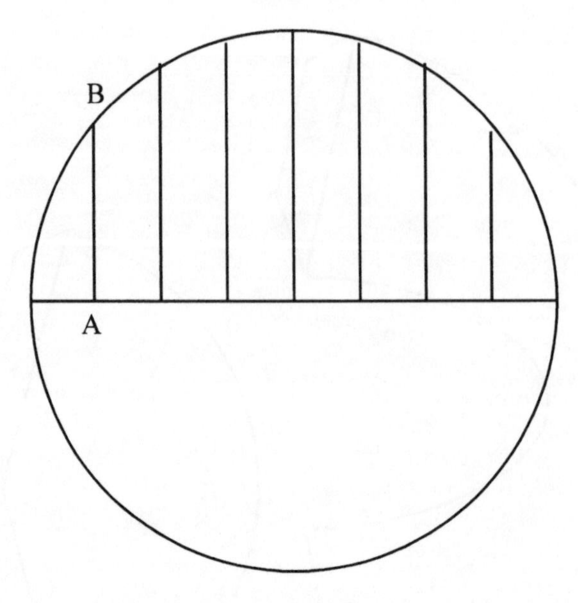

Figure 2

This should be an ellipse, and this way of drawing it shows how it derives from a circle. You do not have to put the dots halfway up the vertical lines; any fraction will do, but it is easiest to start with a half. The circle is called the 'auxiliary circle' of the ellipse.

Later on we will look at a more popular way of drawing an ellipse.

Here in figure 3 is another sight of an ellipse (actually an ellipse and a half!):

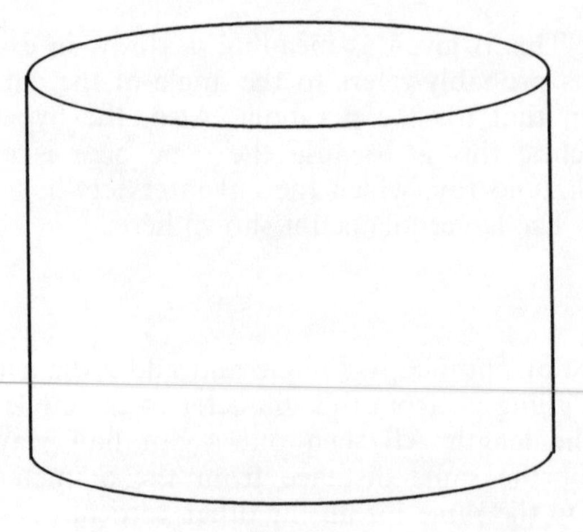

Figure 3

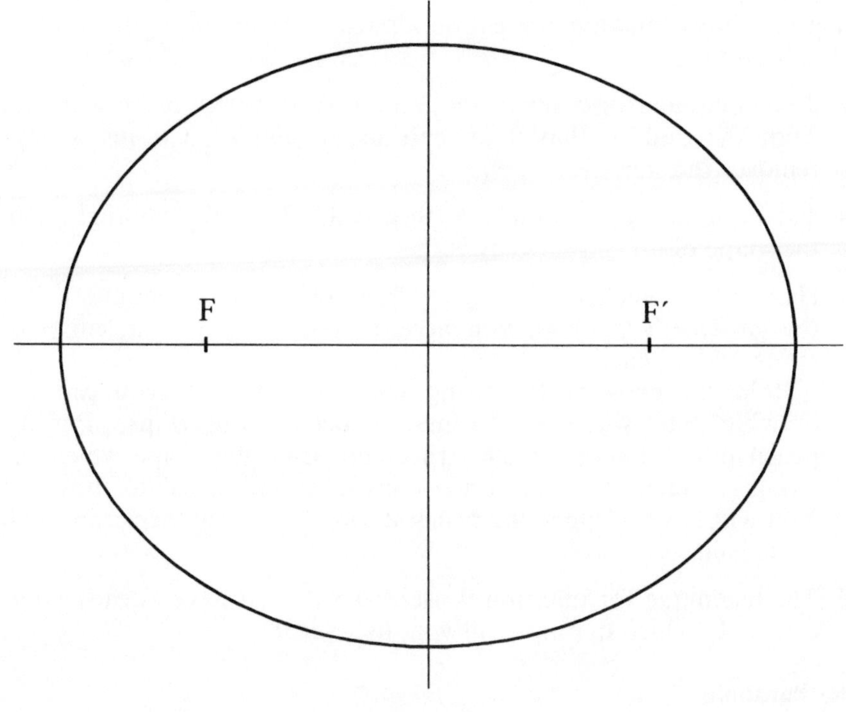

Figure 4

3. Take any point on the ellipse of figure 4 and measure its distance from the point F. Now measure its distance from F′ and add up the two distances. Do the same for more points on the ellipse until you see and confirm the simple result.

The point F and the point F′ are each a focus (plural *foci*) of this ellipse. The ellipse will change its shape if the foci change position.

4. Measure the width of this ellipse, the distance from the left extreme edge to the right edge. (This is called the major axis, the height being the minor axis.) What has this question got to do with question 3?

5. What do you think the ellipse will look like if the foci move so close together they coincide?

6. What do you think the ellipse will look like if the foci move further apart?

Satellites go around the Earth in elliptical paths with the Earth at one of the foci. The foci are usually very much closer together than in the diagram above, making the ellipse almost circular, or in some cases actually circular.

7. Take a circular disc, say a coin, and look at it with one eye closed. Turn the disc so that it appears as an ellipse. (Closing an eye removes the stereoscopic effect.)

8. The pool of light cast on to a (theatrical) stage by a spotlight is in the shape of an ellipse. Why is this?

9. Here is the 'mechanical' way to draw an ellipse, sometimes called the gardener's method: you need a piece of paper, a length of string and a pencil.

Hold the ends of the string with your fingers (you can use drawing pins) where you want the foci of your ellipse. Put the pencil into the loop of the string and draw the shape which the string – which must be taut – constrains the pencil to move in. You will have to move the pencil in order to draw the second half of the ellipse.

10. The technique for question 9 uses a result you have already come across. In which question did you discover it?

The Parabola

WARNING: The work of questions 11 to 17 is rather involved, so you may like to refer to a completed version of the diagram in the answers section.

11. On a photocopy or tracing of figure 5, put a dot at each of these points:

where $y = 4$ and $x = 4$, the point (4,4); where $y = 3$ and $x = 2.25$, the point (2.25,3) (the line where $x = 2.25$ has been drawn for you); the point (1,2); the point (0.25,1) (the line where $x = 0.25$ has been drawn for you); the point (0,0), then at the remaining points which reflect those you have drawn: (0.25,–1); (1,–2); (2.25,–3) and (4,–4).

12. Join the dots with the smoothest curve you can.

The curve you have drawn is a parabola. You will now use this diagram to explore one of its uses. The diagram may become rather messy, so if you draw the lines in answer to each question in a different colour, it may help to keep things clearer.

13. You are now going to draw tangents to this parabola at six different points. A tangent is a line which touches the curve (Latin

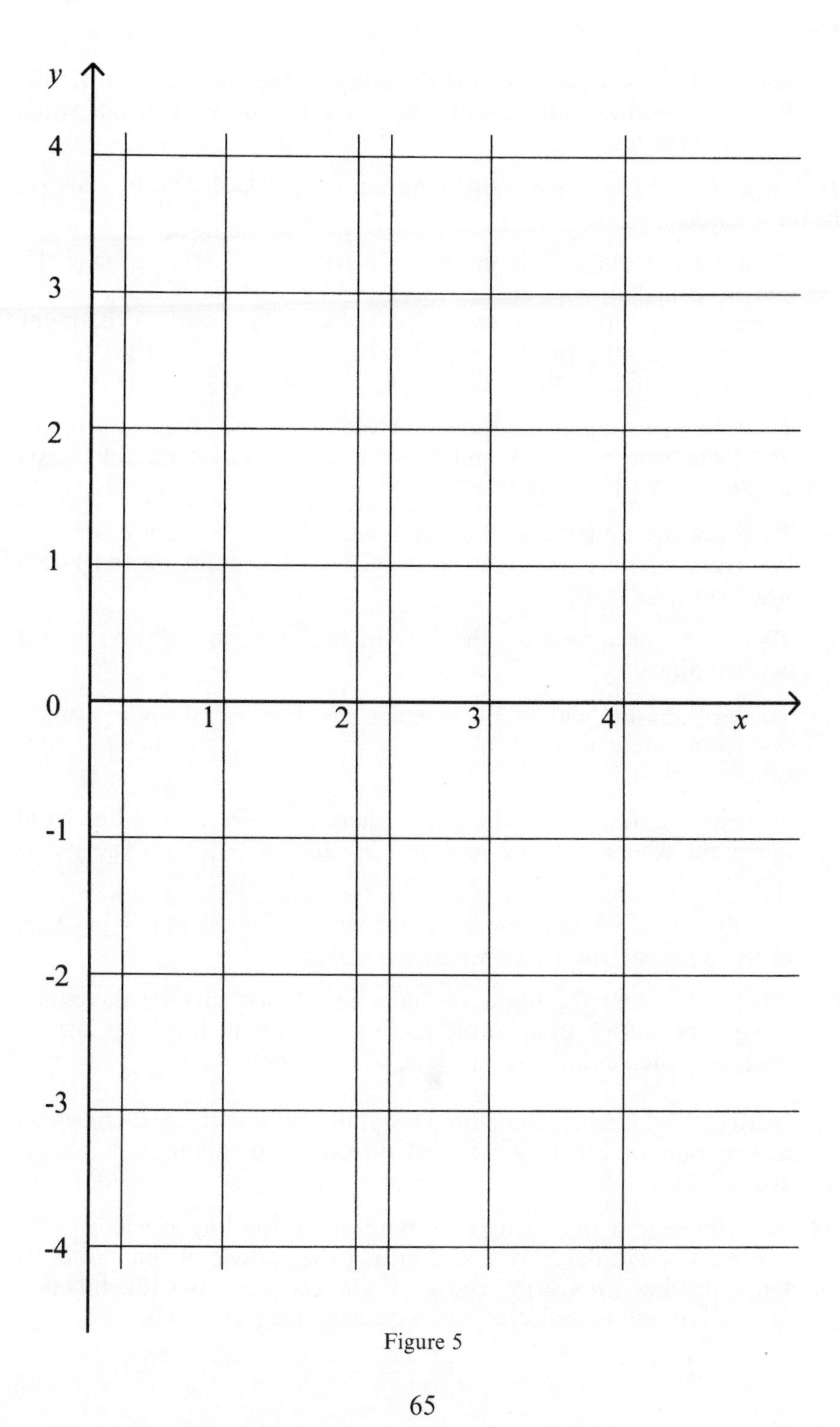

Figure 5

tangere: to touch); it gives the direction of the curve at that point. It is easy to draw for a parabola, but this method will not work for other curves.

It may help you to see what this diagram should look like by looking at the answers.

To draw the tangent at the point where $y = 3$, put your (red?) pen or pencil on the curve where $y = 3$ and $x = 2.25$, then – using a ruler – draw the straight line from this point to the point on the y-axis (the heavy vertical line) where $y = 1.5$ (this is half 3!). You should see that this touches the curve at (2.25, 3).

To draw the tangent at the point where $y = 2$, draw a straight line from where $y = 2$ and $x = 1$, to the point on the y-axis where $y = 1$ (this is half 2).

To draw the tangent at the point where $y = 1$, draw a straight line from where $y = 1$ and $x = 0.25$, to the point on the y-axis where $y = 0.5$.

There is no need to draw the tangent to the curve where $y = 0$; it is there already.

To draw the tangent at the point where $y = -1$, draw a straight line from where $y = -1$ and $x = 0.25$, to the point on the y-axis where $y = -0.5$.

To draw the tangent at the point where $y = -2$, draw a line from the point where $y = -2$ and $x = 1$, to the point on the y-axis where $y = -1$.

Finally, draw a line from the point where $y = -3$ and $x = 2.25$, to the point on the y-axis where $y = -1.5$.

You will notice that the lines you have just drawn envelop the curve. Curves can be shown by drawing lots of tangents in this way; the set of tangents is then called the envelope of the curve.

In a perfect world, when something rebounds off a surface, it comes off at the same angle, but in a different direction, to the angle it hit the surface (see Figure 6).

14. Now imagine a ray of light coming along the line where $y = 2$, two lines above the x-axis, and hitting the surface of the parabola. Draw this line (in green?) and see if you can work out the direction in which it will be reflected. (It is quite an easy answer!)

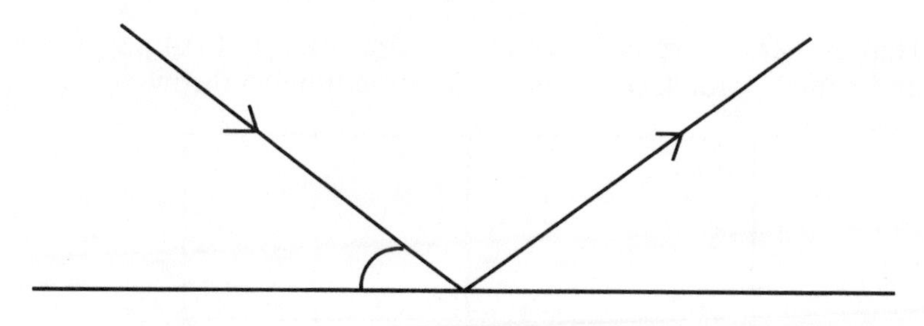

Figure 6

15. Now draw a ray of light coming into the parabola along the line where $y = -2$, two lines below the x-axis. Where will this ray of light be reflected?

16. A ray of light coming along the x-axis and hitting the parabola will be reflected back along its original path. Where do the reflections of questions 14 and 15 meet?

The answer to the last question should be the point (1,0), which is the point where $x = 1$ and $y = 0$.

17. Check, with a protractor if you want to, that rays of light coming into the parabola along the other lines, $y = 3$, $y = 1$, $y = -1$ and $y = -3$, all reflect through the point (1,0).

Note that I have missed out the lines $y = 4$ and $y = -4$ because the diagram would be too cluttered if they were included.

18. What can you make of this result? Read on!

Any rays of light, television signals, sounds etc. which hit a parabolic reflector head-on are reflected to one point, which is the focus of the parabola.

Satellite receivers are parabolic with the pick-up at the focus. Long-distance microphones have parabolic reflectors to concentrate sounds so that conversations can be heard over a considerable distance. The Olympic flame is lit from the Sun's rays after reflection in a parabola has concentrated them at a point.

The idea of reflection is reversible, so that if a source of light is placed at the focus of a parabola, light will be reflected in a narrow beam away from it. This is, of course, how a searchlight works, or a torch. Some electric fires have an element at the focus to direct heat forwards instead of its natural direction of upwards.

It is because all light, sound, etc. comes together at one point that this point is called the focus.

19. Here is a crazy project which needs four pieces of A4 paper, a ruler and pencil, some tape or glue and something to throw.

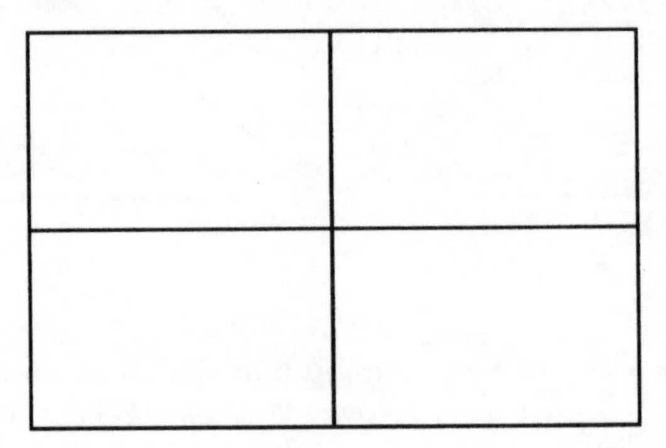

Figure 7

Tape or glue four pieces of A4 paper together in landscape (or horizontal) format to make a larger rectangle the same shape but four times the area, as in figure 7.

Draw a line along the bottom – perhaps a short way up the pages – to act as an axis, and mark the left-hand end of it with a '0'. Then put marks along it every 5 centimetres, marking them 5, 10, 15, ... all the way up to 50.

Now mark 11 points on the paper as follows: the first point is at (0,0); this is where you marked '0' on the line, and 0cm above it. That means it is on the line.

Now mark the point (5, 9); go 5cm along the line to where you marked '5' and put a dot 9cm above it.

Now mark the point (10, 16); go another 5cm along the line to where you marked '10' and put a dot 16cm above it.

Mark the point (15, 21) which is 21cm above the point marked '15'.

In a similar way, mark these points: (20, 24), (25, 25), (30, 24), (35, 21), (40, 16), (45, 9), (50, 0).

Join the points together with the smoothest curve you can, then stick your wonderful graph, which is a parabola, of course, on to a wall so that the first line you drew is horizontal.

The curve shows the path of something which is thrown. Projectiles, as mathematicians like to call things which are thrown, follow the path of a parabola if you ignore such things as air resistance – a reasonable thing to do for small distances such as those here.

Now practise throwing something until it follows the path of your curve. Change the speed of throw and the angle of throw until you manage to do so.

If you put a bucket just below the point (50, 0) you should be able the land your projectile in it, if that's what you really want to do!

20. Figure 8 is a picture of a suspension bridge with the main cable missing. Trace or photocopy this diagram then join the tops of all the vertical lines to complete the picture of the suspension bridge. The main support cable is in the shape of a parabola.

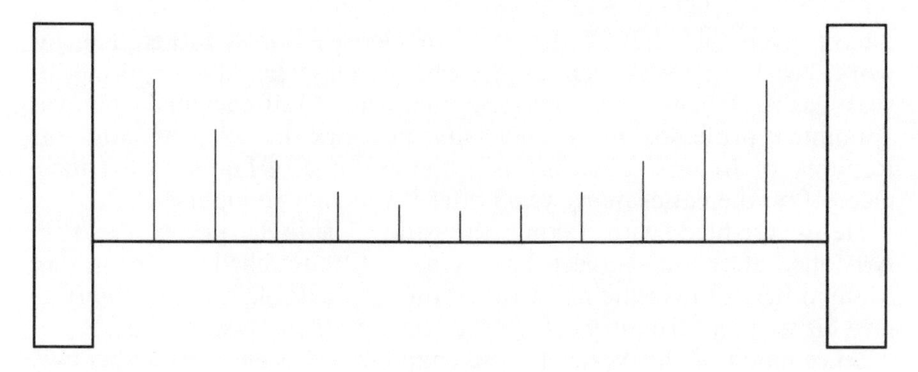

Figure 8

Finally:

One other conic section, the hyperbola, has been mentioned already, although it has few everyday uses.

A further section of a double cone is slightly surprising until you think of it: a pair of straight lines, formed by cutting a double cone in a plane which includes its axis of symmetry. A variable pair of straight lines may have some uses such as cutting the hedge, but not much more.

11

George Boole

For this chapter you will need:

Pen and paper

I have given this chapter the name of George Boole, rather than his work, because he was such an interesting character. After working for years as a village schoolmaster, completely self-educated, he was appointed professor at a university in Cork in what is now the Republic of Ireland. His work is a fine example of finding a use for a piece of mathematics many years after it was first thought of.

He is credited with taking the study of logic beyond that of Aristotle, after the subject had remained untouched for centuries. Many others of his time had worked on it, but Boole had the ability to give his work a formal structure and some mathematical rigour.

Since much of the work of these chapters has been kept deliberately straightforward, I have not gone into anything in any depth. You will therefore get a brief sample of where Boole's work is in use today.

The first examples show how Venn diagrams, which give you pictures of Boole's ideas, are used to solve simple relational problems. It is possible, though not very easy, to develop these ideas to the analysis of complex sentences by first converting statements into symbols then analysing the symbols. If you can manage to do it, the techniques involved can help you to understand legal mumbo-jumbo without employing a solicitor! I wish I could.

Logic gates have developed from Boolean Algebra and can be used as part of various switching circuits. Truth tables help us to understand such circuits. As mathematicians, we do not get our hands dirty by wiring up batteries and bulbs, but the theory of it all is interesting!

I have also dared to introduce some of the symbols of Boolean Algebra because you may like to see how different it is from the algebra you meet at school. It is not quite the same as Boole's original notation but it preserves his ideas.

This chapter is in several sections because Boole's work has many applications. There are also a few difficult questions here and there, many of which are answered in the text shortly after they have been posed. The moral of this little story is that if you find the work hard going, just move on and you may find it becomes easier again. Also, don't be too proud to use the answers; they are there to help.

1. Ten members of Ferndown U3A attend the mathematics course and 15 attend the French course. (a) What is the largest number that could attend the two courses? (b) What is the smallest number that could attend the two courses? (c) If there were only 20 members of the Ferndown U3A, how would this alter your answers to parts (a) and (b) of this question?

An English logician, John Venn, had the bright idea of using diagrams to help answer problems such as question 1. They are called Venn diagrams – not surprisingly.

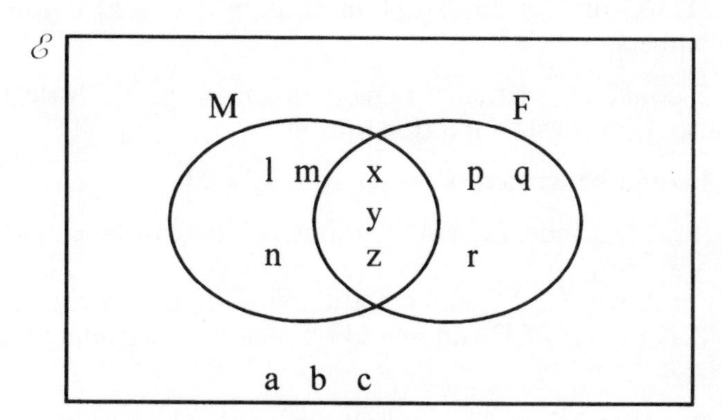

Figure 1

The Universal Set is contained within the rectangle. It is denoted by ℰ, and consists of all the things – people in the case of question 1 – who we consider in the problem.

2. How would you classify the members of the Universal Set for question 1?

The oval labelled M inside the rectangle of figure 1 contains a list of the people attending the maths course; there are only six to save space. m could stand for Mary, z for Zena etc.

3. What letters represent people attending the maths course? (There are not 10.)

The letters inside the oval labelled F stand for the people who attend the French course.

4. What letters represent the people attending the French course?
5. How would you classify the people represented by x, y and z?
6. How would you classify the people represented by l, m, n, p, q, r, x, y and z?
7. How would you classify the people represented by a, b and c?

Although I am doing my best to avoid algebra, it is interesting to see what Boolean Algebra looks like. It is not quite what George Boole used in his time, but it is not much like most people's idea of algebra. The algebra we are most familiar with is arithmetic algebra – algebra which follows the rules of arithmetic.

The answer to question 2 could be written as: $\mathcal{E}$ = {All members of Ferndown U3A} or $\mathcal{E}$ = {a, b, c, l, m, n, p, q, r, x, y, z} if you wanted a list of members.

Question 3 could be written: M = {l, m, n, x, y, z}. Note that the question asked for a list, not a description.

Question 4 could be written: F = {p, q, r, x, y, z}.

The curly brackets indicate a set – things which can be classified in a particular way. So:

M = {Ferndown U3A members attending the maths course} would be read: 'M is the set of Ferndown U3A members attending the maths course'.

If m is a member of set M, we write $m \in M$ which is read: 'm is a member of the set M'.

The answer to question 5 consists of a list of people who attend both the maths and French courses. In the symbols of Boolean Algebra this is $M \cap F$ = {x, y, z} which is read: 'The *intersection* of set M and set F is x, y and z'.

The symbol $\cap$ denotes the intersection of two sets (I think of it as a bridge) and can translate into the word 'and'.

The answer to question 6 consists of people who attend either the maths course or the French course, or both. In Boolean Algebra symbols this is $M \cup F$ = {l, m, n, p, q, r, x, y, z} which is read 'The *union* of set M and set F is l, m, n, p, q, r, x, y and z'. The symbol $\cup$ denotes the union of two sets (It is not a U for union, but it is very much like one!) and can translate into the word 'or'.

The answer to question 7 consists of people who are in the Ferndown U3A but do not attend either the maths or French courses. In symbols this is $(M \cup F)'$ which is read 'not in set M or set F'. The dash represents 'not' (or complement), so M' represents people who are in Ferndown U3A but do not attend the maths course.

Note: Closely related to Boolean Algebra is a study called propositional calculus in which slightly different symbols are used. For example, 'not in M' would be written as $\sim M$ which was used in the chapter on probability.

It could happen that no one attends both maths and French, in which case the intersection of the sets, $M \cap F$, would be empty. An empty set is denoted by $\varnothing$, so we would write $M \cap F = \varnothing$.

Note that we do not write 0; this could be a set with just 0 in it!

Here are some problems. Use Venn diagrams to sort out the muddle of the words. In question 8 there are two sets, so you will need two circles inside your rectangle as in figure 2(a); in question 9 there are three sets, so a Venn diagram like figure 2(b) will be needed.

If you are brave enough, you can practise the use of some Boolean Algebra.

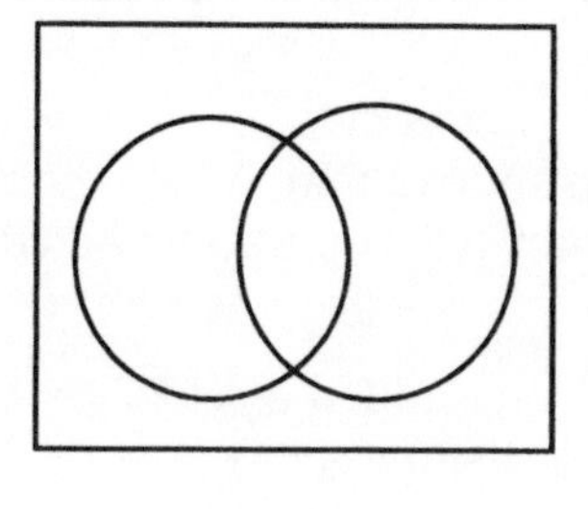

Figure 2(a)

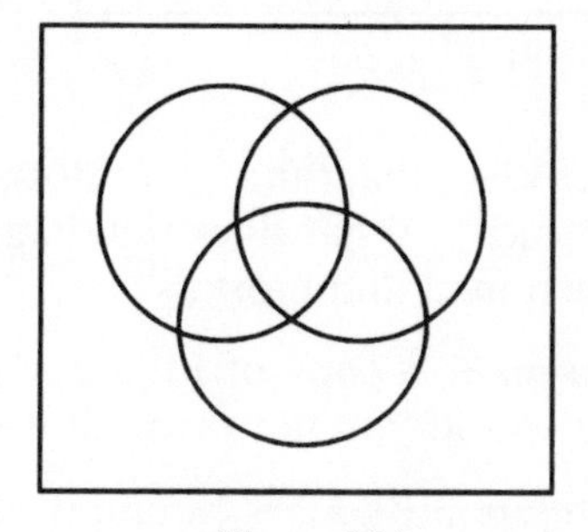

Figure 2(b)

8. There are 25 members of the Brabourne Gardening Club who are all active gardeners. 17 grow flowers and 13 grow vegetables. Use a Venn diagram like figure 2(a) to find out how many grow: (a) both flowers and vegetables; (b) flowers but not vegetables; (c) vegetables but not flowers.

9 The Venn diagram for this question must have three intersecting sets as in figure 2(b). This question is typical of its kind in that you may be given more information than you need, and not all the information is given to you in the order in which it is best used. First fill in the regions in the middle of the Venn diagram, then move outwards. Write in each region the number of people in that set, not a list of names or of letters. Label the sets M, C and S for medium format, compact and single lens reflex (SLR) camera users, respectively.

There are 31 members of the Tricketts Cross Camera Club. Two of them use medium format or compact cameras but not SLRs; five use medium format or SLRs but not compact cameras; of those who do not use medium format, no one uses a compact and SLR; old Joe uses all three types of camera! A total of 21 use medium format, a total of 10 use SLRs and 3 use compact cameras only. Fill in the Venn diagram and find out how many do not use any of these cameras.

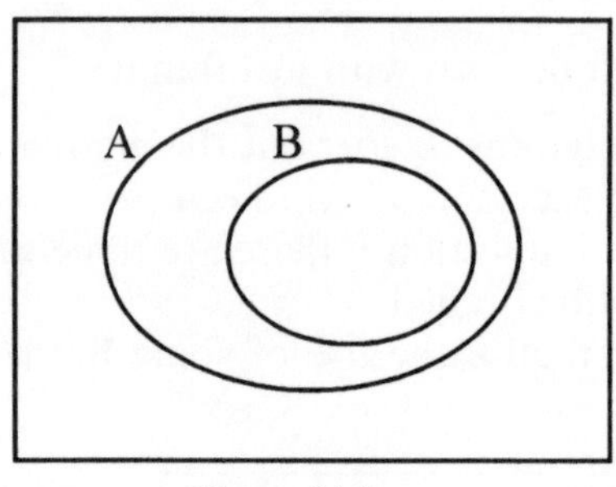

Figure 3(a)

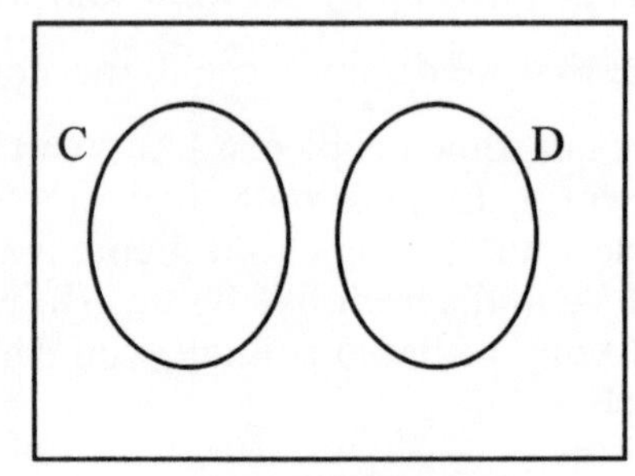

Figure 3(b)

10. In the Venn diagram figure 3(a), suppose that A = {mortals} and B = {men}. What does the diagram tell you about the connection between men and mortals?

11. If you made a copy of Figure 3(a) and added a set W = {women}, where would you (and wouldn't you) show set W?

12. If in figure 3(b), C = {animals of the cat family}, what could set D be?

13. In figure 3(b), what could D not be?

14. Draw four Venn diagrams like Figure 4(a).

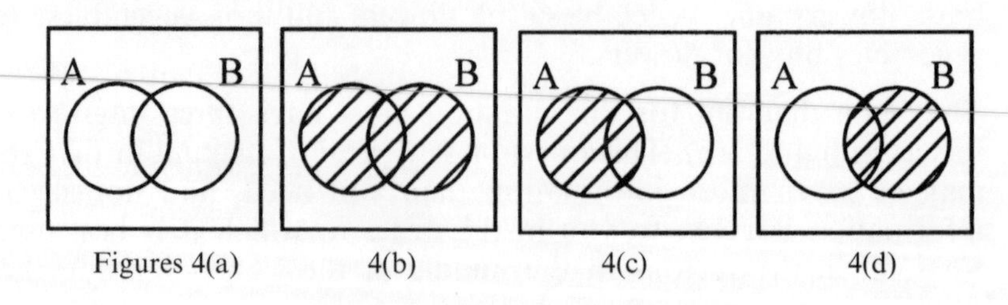

Figures 4(a) 4(b) 4(c) 4(d)

15. The shaded region on figure 4(b) is the set 'in A or B or both'. In symbols this is $A \cup B$. On your first diagram shade in the region which is *not* in A or B or both. In symbols, this is $(A \cup B)'$.

16. The shaded region of figure 4(c) is the set A. On your second diagram, shade in the region which is not A.

17. What is 'Not in A' in symbols?

18. The shaded region in figure 4(d) is the set B. On your third diagram, shade in the region which is not B.

19. What is 'Not in B' in symbols?

20. On your fourth diagram, shade in the region 'Not in A and not in B'. To do this, imagine you can put your second diagram on top of your third. 'Not in A and not in B' will be the region which is shaded twice.

21. What is 'Not in A and not in B' in symbols?

If you have managed this correctly, your first and fourth diagrams should be the same. This means that 'Not in A or B' is the same as 'Not in A and not in B'.
In symbols, this is $(A \cup B)' = A' \cap B'$.

This result is one of deMorgan's Laws. Another is
$(A \cap B)' = A' \cup B'$.

Now let's translate questions 15 to 21 into English!

Inspector Sniffer and his right-hand man Sergeant Quick are investigating a murder at the vicarage. Inspector Sniffer questions Ann and Brian and discovers that they were both out shopping together at the time of the murder. Did either of them commit the murder?

22. What was Inspector Sniffer's conclusion? 'The murderer was...'; (see question 15). What is this in symbols?

Sergeant Quick questions Ann and discovers that she was out shopping.

23. What was Sergeant Quick's conclusion about Ann's guilt? 'The murderer was...'; (see question 16). What is this in symbols?

Sergeant Quick also discovers that Brian was out shopping at the time of the murder.

24. What was Sergeant Quick's conclusion about Brian's guilt? – and in symbols?

When Inspector Sniffer and Sergeant Quick come together to compare notes, Inspector Sniffer says: 'The murderer was – not Ann or Brian.'

Sergeant Quick says: 'deMorgan's Laws tell me the same: it was – not Ann and it was not Brian.'

It is not so easy to make up a silly story about the other one of deMorgan's Laws, but you are welcome to try!

The examples so far have been designed to show that words can be translated into symbols, then reorganised using Boolean Algebra, Venn diagrams and deMorgan's Laws. Here, though, they are inevitably very simple examples.

Another demonstration of the ways in which Boole's ideas have been developed is in switching circuits or logic gates.

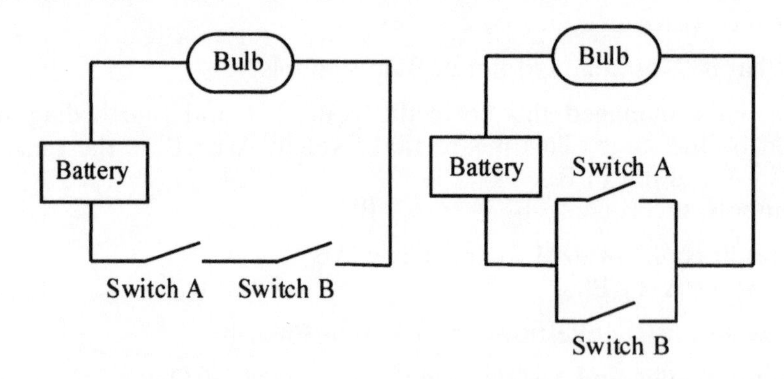

Figure 5(a) Figure 5(b)

The two simple electrical circuits in figure 5 consist of a battery, a bulb and two switches. For the technically minded, figure 5(a) has the switches in series and figure 5(b) has them in parallel.

25. In figure 5(a), will the bulb light up if (a) only switch A is on; (b) only switch B is on; (c) both switches, A and B, are on?

26. In figure 5(b), will the bulb light up if (a) only switch A is on; (b) only switch B is on; (c) both switches, A and B, are on?

The arrangement of switches in figure 5(a) is called an And Gate, and that of figure 5(b) an Or Gate.

It is also possible to have an Exclusive Or Gate through which electricity will flow if switch A or switch B is on, but not if both are on.

Another kind is a Not Gate, where current will flow if the switch is off, but not if it is on. An arrangement something like this is used in most burglar alarms where the siren or bell will sound if the current is switched off, thus protecting against wires being cut.

The effects of figures 5 (a) and (b) above can be summarised in Truth Tables. They use a system like my electric kettle where '1' means 'on' and '0' means 'off'.

Truth Table for an And Gate

A	B	Bulb
0	0	0
1	0	0
0	1	0
1	1	1

Truth Table for an Or Gate

A	B	Bulb
0	0	0
1	0	1
0	1	1
1	1	1

The first column of each table shows the state of switch A, and the second column the state of switch B. The third column shows the state of the bulb.

The first row of each table shows 0 0 0, which means 'switch A is off, switch B is off and the bulb is off'.

The second row shows A on and B off; for the And Gate the bulb is off but for the Or gate it is on.

The third row shows much the same information except that A is off and B is on.

The fourth row shows that when both switches are on, the bulb is on, too.

The first two columns show all the possible arrangements of the switches A and B.

27. See if you can draw up a truth table for an Exclusive Or Gate. It should have two switches A and B as input in the first two columns, showing all possible arrangements of these switches, like the tables above, and a bulb in the third column. Or course, these logic gates can do a lot more than light up bulbs, but this will do for our purposes.

28. Draw another truth table (with just switch A and a bulb) for a Not Gate.

Here is a very neat device called a latch which uses an Or Gate. You have almost certainly used one yourself!

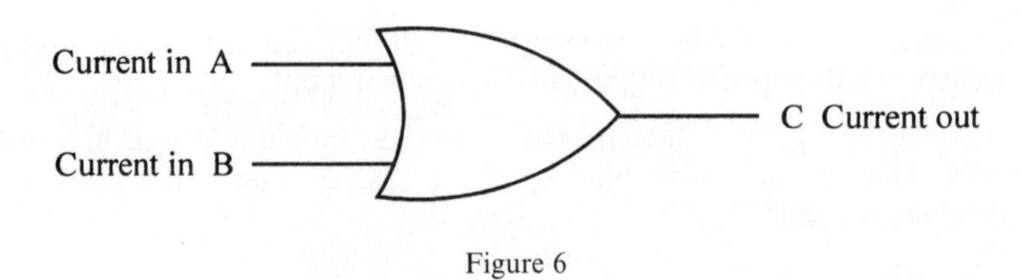

Figure 6

Figure 6 is a diagram of an Or Gate. If an electric current flows into it along either A or B, then current will flow out of C. Figure 7 shows a simple change to an Or Gate where current will flow in only along B.

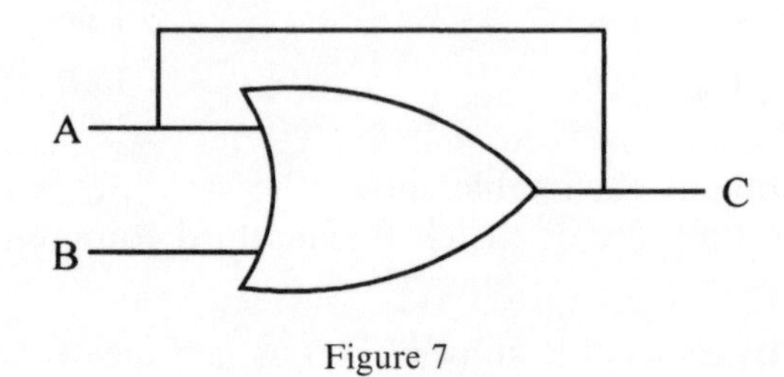

Figure 7

29. If current flows into the Or Gate along B for any length of time, will it flow out along C?

Some extra circuitry has been added to the Or Gate to take current from C back to A.

30. If current flows through C, will it flow in through A?

31. If current is flowing in along A, will it flow out through C?

To summarise the answers to the last few questions, if someone presses a button which sends an electric current through B, it will flow through C. If it is then re-routed back through A, it will continue to flow through C whether it is still flowing through B or not.

You use a latch when you press the button on a pedestrian-controlled crossing. Once you press the button, the machinery knows you are there and will stop the traffic eventually. People who impatiently press the button a dozen more times are wasting their time and energy.

We have looked at some simple examples of applications of George Boole's ideas. I am sure he had no idea where his thoughts would lead to, but I always think he should be an icon for people who have no qualifications. The lack of a university degree did not stop him contributing to mathematical knowledge or achieving status in the academic world. I am not sure it could happen today, but you never know.

Trivial Pursuit!

George Boole died at the age of 49. The story goes that he died of a chill after getting wet on his way to a lecture. It does rain in Ireland! He left a widow and five daughters who all made something of their lives – a notable achievement in Victorian times.

One of his daughters became a professor of chemistry, and another, Ethel Lillian, married a Polish scientist called Wilfrid Voynich and made a name for herself as a novelist. Her most famous novel, *The Gadfly*, was very popular in Russia, where three operas and a film were based on it. Dmitri Shostakovich's music for the film is still played from time to time.

Some of the information for this section came from Martin Gardner's book *Mathematical Circus*, Chapter 8.

12

Some Areas and Volumes

For this chapter you will need:

Pen and paper
A calculator
A photocopy of figure 1
A photocopy of figure 2
3 photocopies of figure 3

This chapter deals with areas and volumes of circular things and cubical things. Where circular things are involved, we inevitably come up against π. This is another one of that special set of numbers which cannot really be found, sometimes called transcendental numbers or irrational numbers. The number we met in chapter 1 on the Golden Section was another example. Take $\pi = 3.142$.

I hope you will take time to cut out and glue the shapes. I still find the comparison of the volumes of two cubes something of a surprise, even after years of doing the experiment.

You will certainly know more about small mammals than I do, but it is still interesting to see why there is a lower limit to their size.

1. Make up a cube which has all its edges 4cm long. The net is figure 1 of this chapter.
2. Make up a cube which has all its edges 2cm long. The net is figure 2 of this chapter.
3. Look at the 4cm cube: each face is a square, and each side of this square is 4cm long. What is the area of this square – one face of the cube?
4. Remembering that the cube has 6 faces, what is the total surface area of the 4cm cube? (What is the area of all 6 faces added together?)
5. Look at the 2cm cube: each face is a square of side 2cm. What is the area of one face?

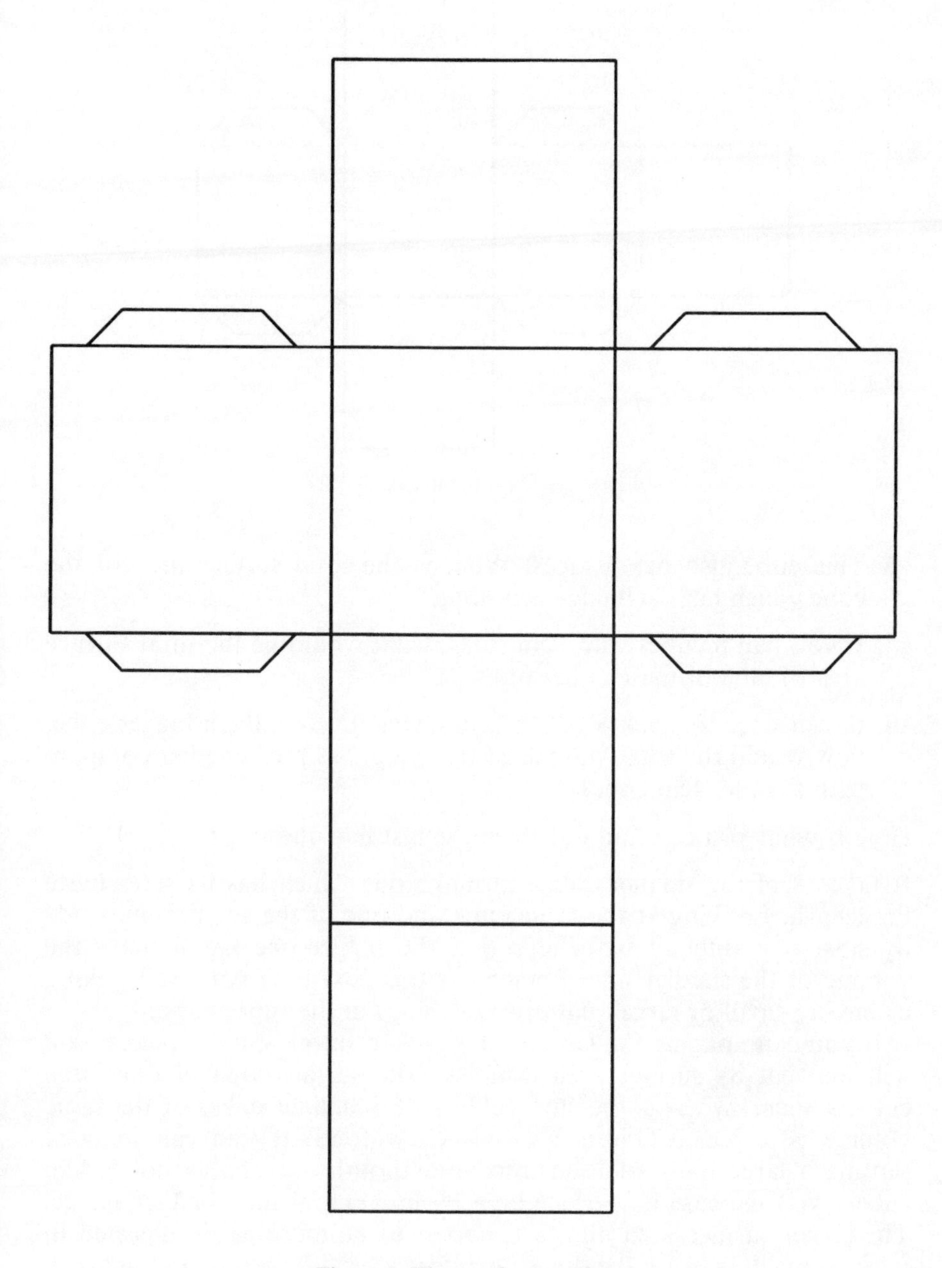

Figure 1: The net for question 1

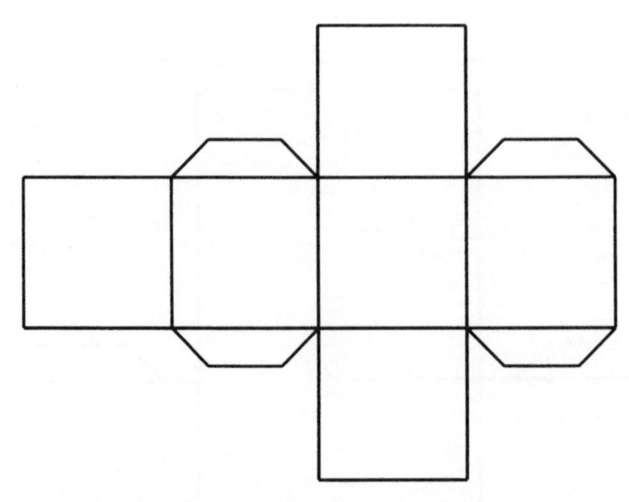

Figure 2: The net for question 2

6. This cube also has 6 faces! What is the total surface area of the cube which has each edge 2cm long?

7. If you had 8 cubes with 2cm sides, what would be the total surface area of all 8 of these cubes put together?

8. If you were to stack 8 of the 2cm cubes up to make a bigger cube, how would the total volume of these 8 cubes put together compare with the one 4cm cube?

Here is what you can find out from the last few questions:

It takes 8 of the smaller cubes to make one which has its sides twice the length. Looking at the large cube and one of the smaller ones side by side, it is difficult to believe that the bigger one has 8 times the volume of the smaller one. Remember this next time you are thinking of buying small or large quantities of things in the supermarket!

If you can imagine the larger cube broken into 8 smaller pieces, you will see that its surface area doubles. The surface area of one large cube is $96cm^2$ whereas the surface area of 8 smaller cubes of the same volume is $192cm^2$. This is why we chew food! If you can imagine putting a large lump of food into your mouth – probably not a 4cm cube – you increase its surface area by breaking it into smaller pieces. The larger surface area allows the food to be more easily digested in your stomach because the digestive juices have a larger area to act on.

For mathematicians or scientists, my apologies for the next few questions. It is not quite the right thing to compare areas with volumes, muddling up the dimensions, as we are going to, but it does make the point.

9. Consider first the 4cm cube: what is its volume? (Its length × breadth × height.)

10. Its surface area is the answer to question 4. How many times bigger is its surface area than its volume? (Ignore the dimensions!)

11. Consider now one of the 2cm cubes: what is its volume?

12. Its surface area is the answer to question 6. How many times bigger is its surface area than its volume?

13. Consider a cube which has all its edges 1cm long. Each face is a square of side 1cm. What is the area of one face?

14. What is the total surface area of the cube (i.e. all 6 faces together)?

15. What is the volume of the cube?

16. How many times bigger is its surface area than its volume?

You will notice from the last few questions, that as a cube gets smaller, its surface area increases in proportion to its volume. This is why there is a minimum size for a mammal. A mammal loses heat through its skin, the amount of skin depending on its surface area. It produces heat within the volume of its body. Any mammal smaller than a shrew will lose so much heat through its skin, and have such a small volume in proportion, that it cannot generate enough heat to sustain its life.

17. Make three copies of the net of figure 3, then use them to make up three 3-D shapes. The shapes will be identical lopsided square-based pyramids.

18. Here is a 3-D jigsaw puzzle: fit the three shapes together to make a cube.

The outcome of all this work is to show that the volume of a pyramid is one third the area of its base multiplied by its height. This is clear if you realise that each shape is one third of the cube, and that it is also a square-based pyramid whose height is the length of one side of the cube.

The result above will be true if the base of the pyramid changes shape to, say, a circle (in which case the shape will be a cone), and also if the pyramid is not lopsided. (Or even if it is more lopsided.)

19. Find the surface area of a tennis ball. (Find it later – it will be much easier to do then!)

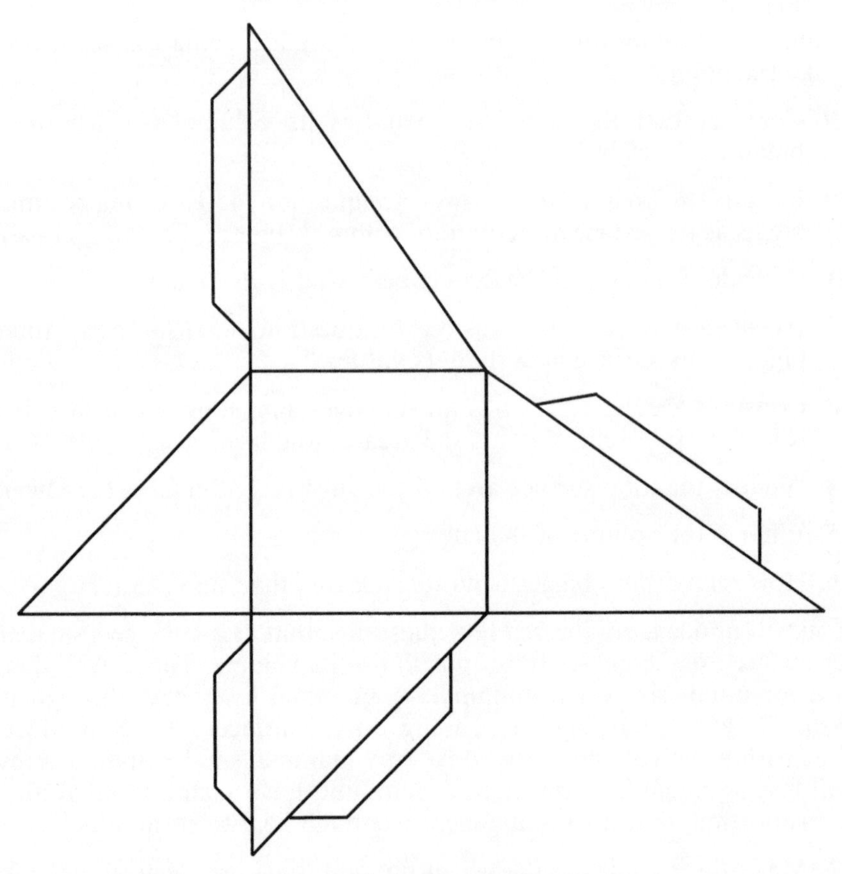

Figure 3: The net for question 17

Figure 4 shows a sphere with a cylinder wrapped around it so that the cylinder is the same height as the sphere, and has the same radius.

The surface area of the sphere is the same as the surface area of the cylinder. In fact, if you were to take the top centimetre, or inch, or fathom, of the sphere, its surface area would be the same as the top centimetre, or inch, or fathom, of the cylinder.

20. Again! Take a tennis ball – or any other sphere – and cut a strip of paper as wide as the ball is high. Then wrap the paper around the ball and mark where the ends overlap.

21. Open out the paper, measure its length and its width, multiply these two lengths together and you have the surface area of the ball.

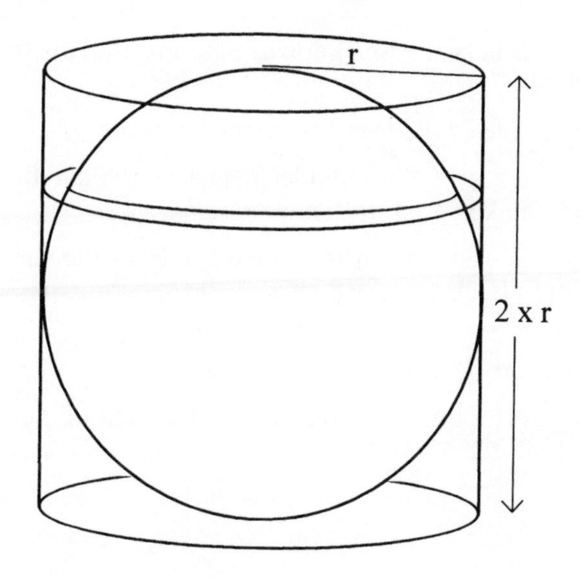

Figure 4

Some Areas And Volumes Extra

Here are two formulas written as you might use them on a calculator:

The circumference of a circle $= 2 \times radius \times \pi =$

The area of a circle $= \pi \times radius \times radius =$

You may have noticed in algebra that letters which are in place of numbers – called 'pronumerals' – are always in italics. I have followed that pattern above, so when you see *radius*, you substitute for it the number which is the length of the radius.

To find a formula to calculate the surface area of a sphere:

22. Remember that the sphere is surrounded by a cylinder. What is the height of this cylinder if we call the radius of the sphere *radius*?

23. The length of the piece of paper which makes up this cylinder is the circumference of a circle of radius *radius*. What is this circumference (use '*radius*' not '*diameter*') ? (Note that you can also measure this length.)

24. The area of the piece of paper which makes up this cylinder is its length × its height. In other words, the area is answer 22 × answer 23. What is this?

To save you a nervous breakdown, here are the answers to questions 22, 23 and 24:

The height of the cylinder is 2 × *radius*.

The circumference of this cylinder, which is the length of the piece of paper which makes it up, is 2 × π × *radius*.

Multiplying the last two answers together gives the surface area of the cylinder, and also the surface area of the sphere. It is 2 × *radius* × 2 × π × *radius*.

This simplifies to 4 × π × *radius* × *radius*.

Note that this is 4 times the formula for the area of a circle. In conventional algebra, it is $4\pi r^2$.

13

Movement

For this chapter you will need:

Pen and paper
Calculator

This chapter could easily be classified as physics if you felt that classification was necessary. Mathematics and physics do intertwine.

Our hero Joe Leggit shows his prowess at high jump, not only here on Earth, but also on a variety of planets and on the Moon. In this way, perhaps you will see something of the effect of gravity.

The final part of the chapter deals with circular motion both on Earth and in space. I am not sure that your driving will improve if you understand something of the forces necessary to make a car go round in a circle, but I can always hope, especially if you are driving on the same road as me but in the opposite direction! The references to space give you some idea of how to keep a communications satellite in place.

I have not considered the extent of the accuracy with which satellites have to be placed in orbit, especially when they have to meet up with a space station. There is an awful lot of space out there.

However, I hope that what I have included gives you food for thought.

Interplanetary High Jump:

We have already met Joe Leggit the sprinter in earlier chapters, but he is also quite a good high-jumper. His extraordinary technique, known in Tricketts Cross as the Leggit Flop, takes him over the bar in a posture which suggests he is lying on a slab in the morgue. Figure 1 shows the speed/time graph of his descent, after he has cleared his personal best height. Note that Joe is falling, even though the graph is rising. The graph is rising because Joe's speed is increasing as he falls.

1. How long does it take Joe to fall to the ground? (Read off the 'Time' axis.)

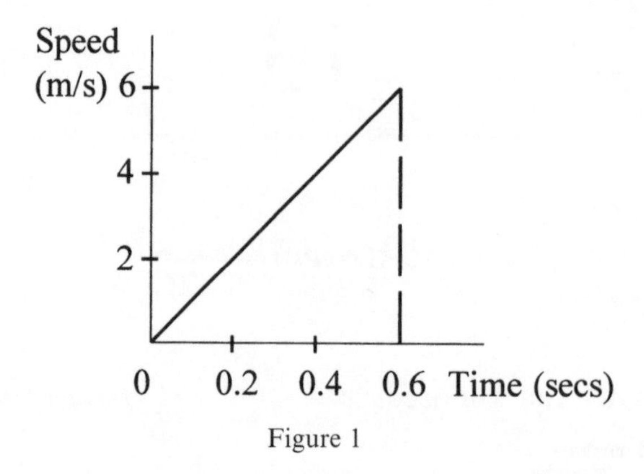

Figure 1

2. How fast is Joe travelling when he hits the ground? (Read off the 'Speed' axis.)

3. Through what distance does he fall? (This is the area under the graph, measured in the units of the graph.)

4. What is Joe's acceleration as he falls? (Or: By how much would his speed increase in 1 second?)

The answers are: Joe falls to the ground in 0.6 seconds, hits the ground at a speed of 6 metres per second and falls (and therefore jumps!) a height of 1.8 metres. Since Joe falls under the influence of gravity, the answer to question 4 shows the acceleration due to gravity. I have taken it to be 10m/s/s although it is slightly less than this. It can also vary according to your distance from the centre of the Earth.

Joe later took up interplanetary travel. Before he left, he used the formula

$$v^2 = 2 \times g \times \text{height}$$

to calculate how fast he propelled himself when he jumped 1.8m. v is his take-off speed and g is the acceleration due to gravity.

5. Calculate the speed Joe took off at if $g = 10$m/s/s, the acceleration due to gravity, and 'height' is 1.8m, the height that Joe could jump.

6. When Joe reached the Moon, he found that the acceleration due to gravity was only 0.1653 its value on Earth. What is the acceleration due to gravity on the Moon? (Hint: It is $0.1653 \times g$, or 0.1653×10 m/s/s.)

7. He knew that the speed he could take off at when he jumped was 6m/s, so its square, v^2, was 36. He rearranged his formula to calcu-

late how high he could jump. It was: height = 18 ÷ gravity. How high did Joe jump?

8. Inspired by his incredible personal best, Joe set off for Venus where he found that the acceleration due to gravity was 0.9 its value on Earth. What is the acceleration due to gravity on Venus, measured in m/s/s ? (Hint: It is 0.9×10 m/s/s.)

9. How high did Joe jump on Venus? (Use the formula of question 7: height = 18 ÷ 9.)

Joe then explored many other planets, and recorded the acceleration due to gravity compared with Earth in the list below:

Mercury: 0.38; Venus: 0.9; Earth: 1; Mars: 0.38; Jupiter: 0.26; Saturn: 1.2; Uranus: 0.9; Neptune: 1.2; Pluto: 0.1.

10. On which planet did Joe achieve his personal best high jump, and what was it?

If you want to calculate how high Joe jumped on all the other planets, you are welcome to do so. The results are listed in the answers section.

Going Round in Circles

11. Tie something to the end of a piece of string and whirr it round in a circle. Does the string stay taut?

12. Provided you are not going to break anything, let go of the string as it is turning. What happens to the thing on the end of the string?

13. Can you see the meaning of the expression 'flying off at a tangent'?

The string stays taut because there is a force along it, its tension, which is constantly pulling the object on the end of it in towards the centre of the circle. This is called the centripetal force, and is necessary to keep the object moving in a circle. When this force is removed, the object at the end of the string flies off at a tangent; in other words it leaves its circular path.

It is possible to calculate the force necessary to keep something moving in a circle.

You must know the mass of the object, in kilograms.
You must know the speed of the object, in metres per second.
You must know the radius of the circle it is moving in, in metres.

To find the force, calculate:
(*Mass of object*) × (*Speed of object*) × (*Speed of object*) = ÷ (*Radius of circle*) =

Note: this formula has been written in the way you would use it on a calculator.

14. What is the force needed to keep a 1000kg (= 1 tonne) car travelling at 13m/s in a circle of radius 50m?

15. What is the force needed if the same car increases its speed to 26m/s around the same bend?

13m/s is more or less 29 miles per hour, or 46.8km/hr. Notice that when the speed doubles, the force necessary to keep the car in a circle increases fourfold. This force is provided by the friction between the tyres and the road, so if the road is slippery, the possibility of skidding is increased considerably with the extra speed.

16. What force is needed to keep a 40000kg (40 tonne) lorry travelling at 13m/s in a circle of radius 50m?

Here the increase in the force is directly proportional to the increase in the weight of the lorry which is why lorry tyres have to be so much heavier than car tyres.

17. What force is necessary to keep a 1000kg car travelling at 13m/s in a circle of radius 25m?

Here the force on the tyres doubles as the radius of the circle halves; this is inverse proportion. This is why a racing driver will try to take corners driving round in a circle with the largest possible radius. In this way, he or she can maintain a good speed with less fear of the force on the tyres increasing to the point where the car skids.

Interplanetary travel in circles

Satellites go around the Earth in elliptical orbits, with the Earth at one focus of the ellipse. The Moon orbits the Earth in this way, and the Earth also orbits the Sun this way, although both of these orbits are almost circular because the two foci of each elliptical path are very close together. Circular orbits are possible.

18. A satellite orbits the Earth at a speed of 3054m/s. How many metres does the satellite travel in one minute?

19. How many metres does it travel in one hour?

20. How many kilometres (1km = 1000m) does the satellite travel in one hour? (Note: this is its speed in kph)

21. This satellite travels in a circular orbit with a radius of 42000km. How far does it travel in one orbit? (Distance = $2 \times \pi \times$ radius)

22. How long does it take this satellite to complete one orbit of the Earth? (This is: answer to question 21 divided by answer to question 20.)

Notice that this satellite will orbit the Earth in the same time that the Earth rotates once. This means that it will remain directly above a particular place on the Earth. A satellite which does this is said to be in a geostatic, or geostationary, orbit. Television and other communications satellites are placed in geostatic orbits.

14

Virtual Machines

For this chapter you will need:

Pen and paper
A calculator if you wish

Virtual machines do not exist; if they did, they would be machines, not virtual machines! The idea behind each of the two used in this chapter is straightforward but can lead to some interesting thoughts.

I am not sure that computer programmers would use flow diagrams these days although popular rumour has it that they do. They are very useful for giving a broad outline of a complicated set of instructions, but they are used in this chapter where they seem to me to be particularly useful: showing calculations where a sequence of operations is repeated. A computer programmer would call such a repeated sequence a 'loop', and the loop structure can be seen physically in the diagram. Again, I hope all this will become much clearer as you sail blissfully through the chapter.

Hint: the work of the Extra section is a repeat of part of chapter 2.

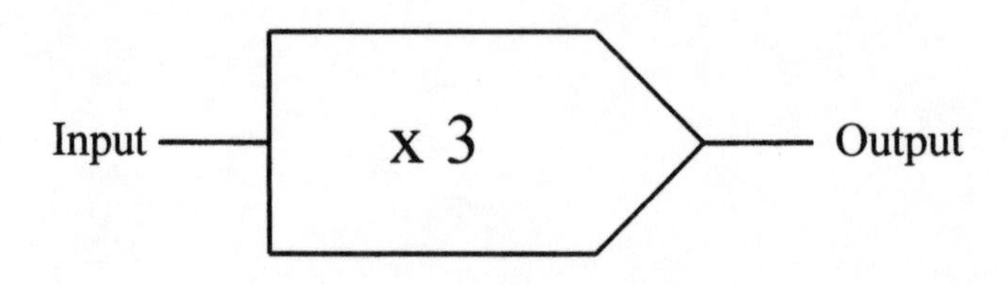

Figure 1

Imagine that you have a machine like the one in figure 1 which will multiply by 3; that's all!

Find the output when the input is:

1. 2 **2**. 5 **3**. 3 **4**. 17

Find the input when the output is:

5. 33 **6**. 51 **7**. 15 **8**. 342

I am sure that at the moment you can see that the machine is a simple way of testing your multiplying and dividing. However, it does emphasise the point that multiplying and dividing are opposites; in mathematics we say they are 'inverse' operations.

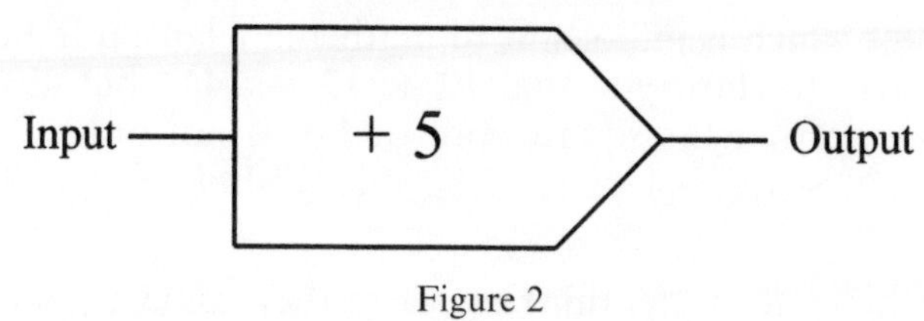

Figure 2

Using the machine in figure 2, find the output when the input is:

9. 6 **10**. 14 **11**. 100 **12**. 195

Find the input when the output is:

13. 11 **14**. 19 **15**. 42 **16**. 100

This machine shows that adding and subtracting are inverse operations.

These machines, virtual machines because they do not exist in the real world, are sometimes called 'function machines' but are better called 'number machines'.

For the machine of figure 3, the word 'square' means 'multiply the number by itself'. If you square 7, you get 49 because $7 \times 7 = 49$.

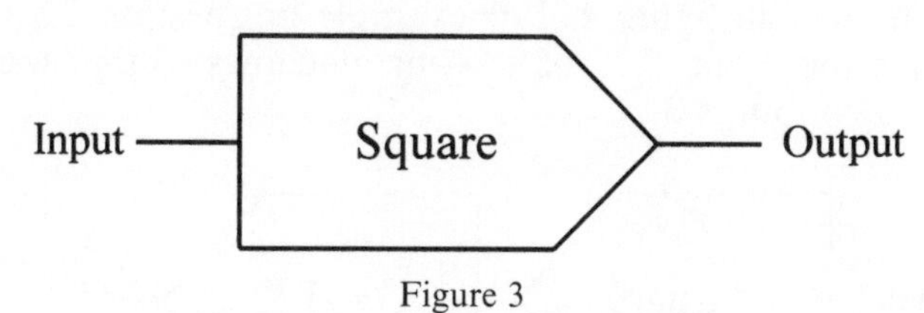

Figure 3

For the machine in figure 3, find the output when the input is:

17. 2 **18**. 6 **19**. 12 **20**. –12*

* You may want to use a calculator here; if so, to enter –12, first enter the number 12, then press the +/– button. If you have no +/– button, –12 can be 0 – 12.

Find the input when the output is:

21. 4 **22**. 81 **23**. 144 **24**. 2 (If you have a calculator.)

The opposite, or inverse, of squaring a number is finding its square root. Things, or people, come from their roots; squares come from their square roots. On a calculator, the square root is shown by the symbol $\sqrt{}$, so $\sqrt{4} = 2$ (or -2).

Here is a point which is important to mathematicians: questions 19, 20 and 23 are there to show you that $12 \times 12 = 144$, but also $(-12) \times (-12) = 144$, so that the square root of 144 is not only 12, it is also -12. We should say $\sqrt{144} = \pm 12$; the symbol $\pm$ is read 'plus or minus'.

Number machines start you thinking when they come in twos.

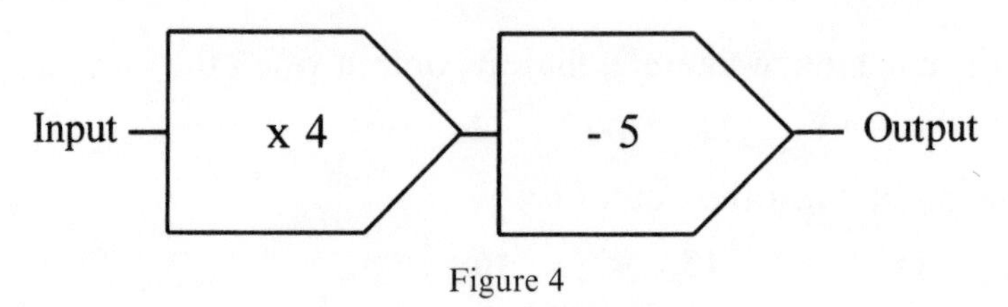

Figure 4

For the machines in figure 4, find the output when the input is:

25. 2 **26**. 7 **27**. 10 **28**. 1

Find the input when the output is:

29. 3 **30**. 15 **31**. 19 **32**. 41

33. Draw a pair of number machines which work the other way round from the ones in figure 4. For example in question 29, the output is 3 when the input is 2; make your machines so that the output is 2 when the input is 3.

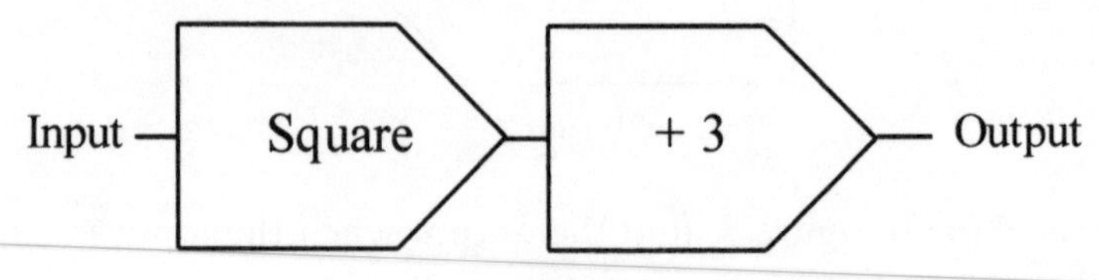

Figure 5

For the machines of figure 5, find the output when the input is:

34. 4 **35**. 5 **36**. –5 **37**. 10

Find the input when the output is:

38. 19 **39**. 7 **40**. 84 **41**. 6 (If you have a calculator.)

Note that the last four questions should have two answers each; for example, question 38 can have an input of 4 or –4. If you use a calculator for question 41, or any of the others, it is unlikely that it will show $\pm$ 1.732... on the display when you ask it for $\sqrt{3}$. You are expected to know all about $\pm$.

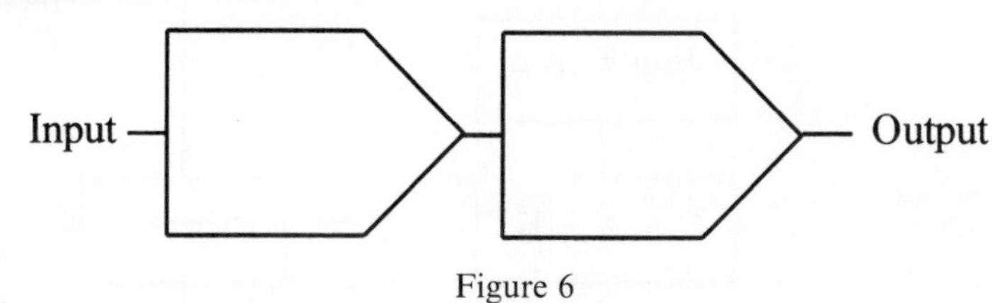

Figure 6

42. If in figure 6, the input is 3 and the output is 21, what operations can you put in the two machines?

43. If in figure 6, the input is 1 and the output is also 1, what operations can you put in the two machines?

There can be many different answers to both question 42 and question 43. Probably the easiest answers to question 43 are pairs of inverse operations, such as: + 2 and –2 or × 17 and ÷ 17.

Flow Diagrams

The flow diagram, such as figure 7, is another virtual machine which is allegedly used by computer programmers. In the simple examples of this chapter, instructions are given in rectangles, and questions requiring a decision are given in diamond-shaped boxes. You must also imagine that stores, rather like calculator memories, can remember numbers for you; these numbers can be altered. The stores are labelled by capital letters. In figure 7, the store A starts with the number 0 in it, the store B has the number 3 in it, and C starts with the number 0. To get information out of the machine, we say ‘Print’; so ‘Print B’ would give you 3, unless the number in store B is altered.

44. Examine the flow diagram, figure 7. Part of it follows a loop. See if you can follow the path of the flow, but do not bother to find out what it all means.

45. There is a simple way to find out what is going on in this virtual machine. You draw up three columns (for this machine), one for

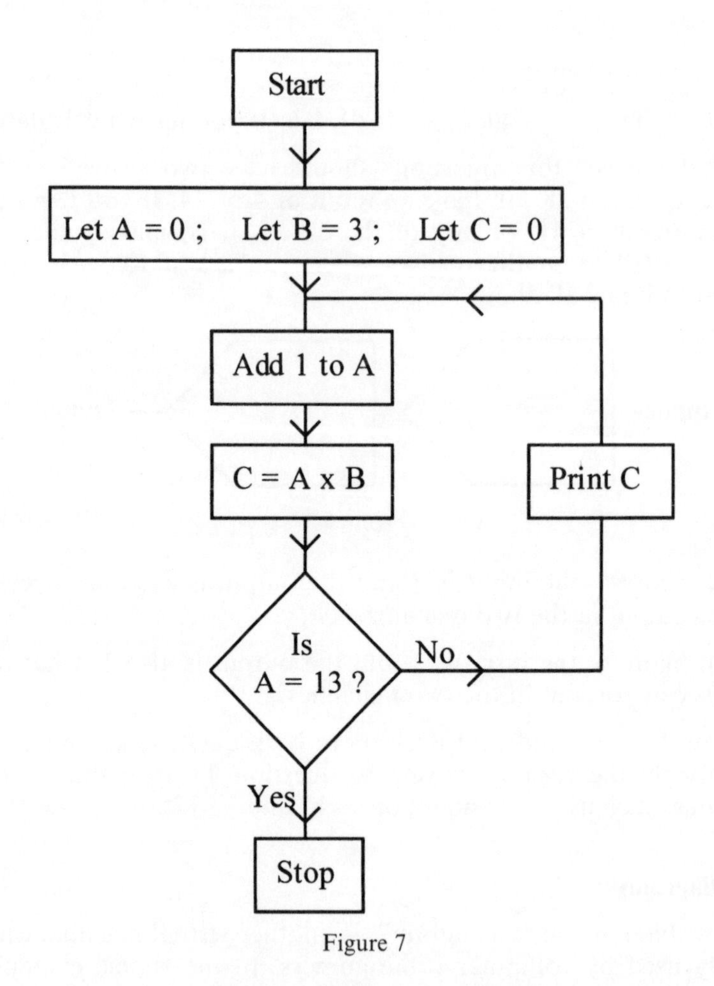

Figure 7

each store. As you go round and round the machine, you write down what numbers are in which store as you go round each loop. The first three entries have been done for you. Copy and complete it:

A	B	C
0	3	0
1	3	3
2	3	6

46. Can you work out what this flow diagram is doing? Do not answer: 'Using a sledge hammer to crack a peanut'! This is an introductory example.

Notice that store A is a counter, store B keeps the same number throughout and store C is the means of printing out the answers.

47. What simple alteration would you make to the flow diagram so that it printed out the answers to the 4 times table?

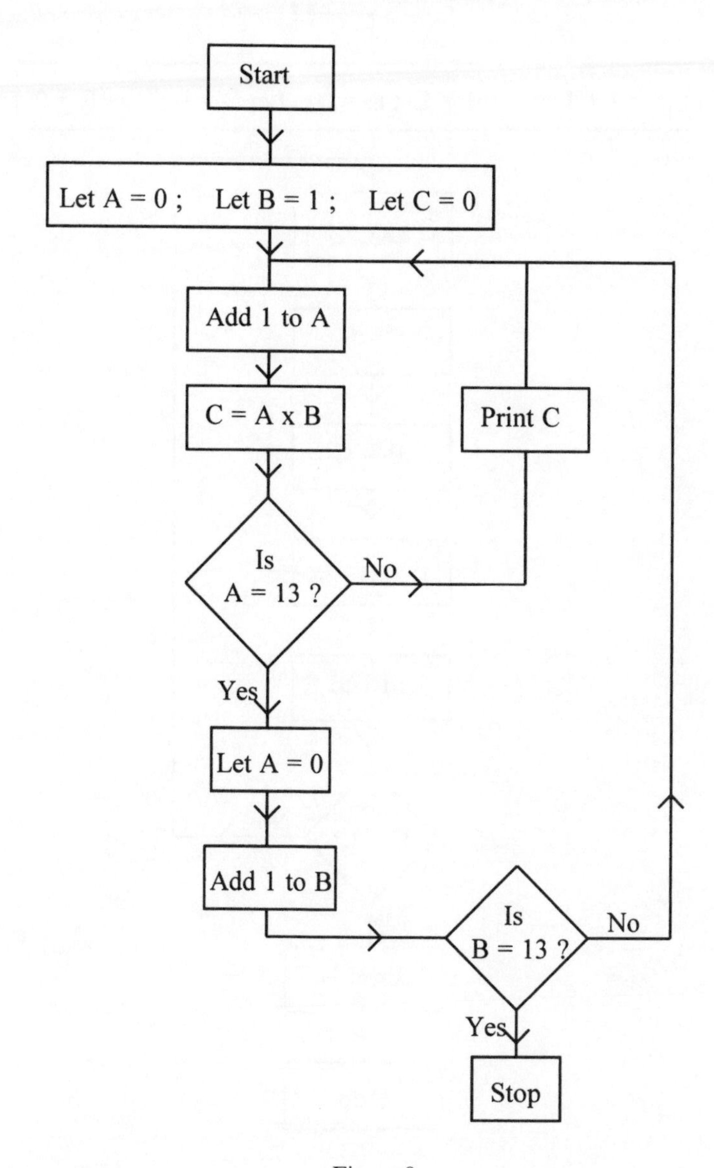

Figure 8

48. Use three columns as in question 45 to find out what the flow diagram of figure 8 is doing. Stop when you have seen what it is doing; if you work all the way through it you will miss your tea!

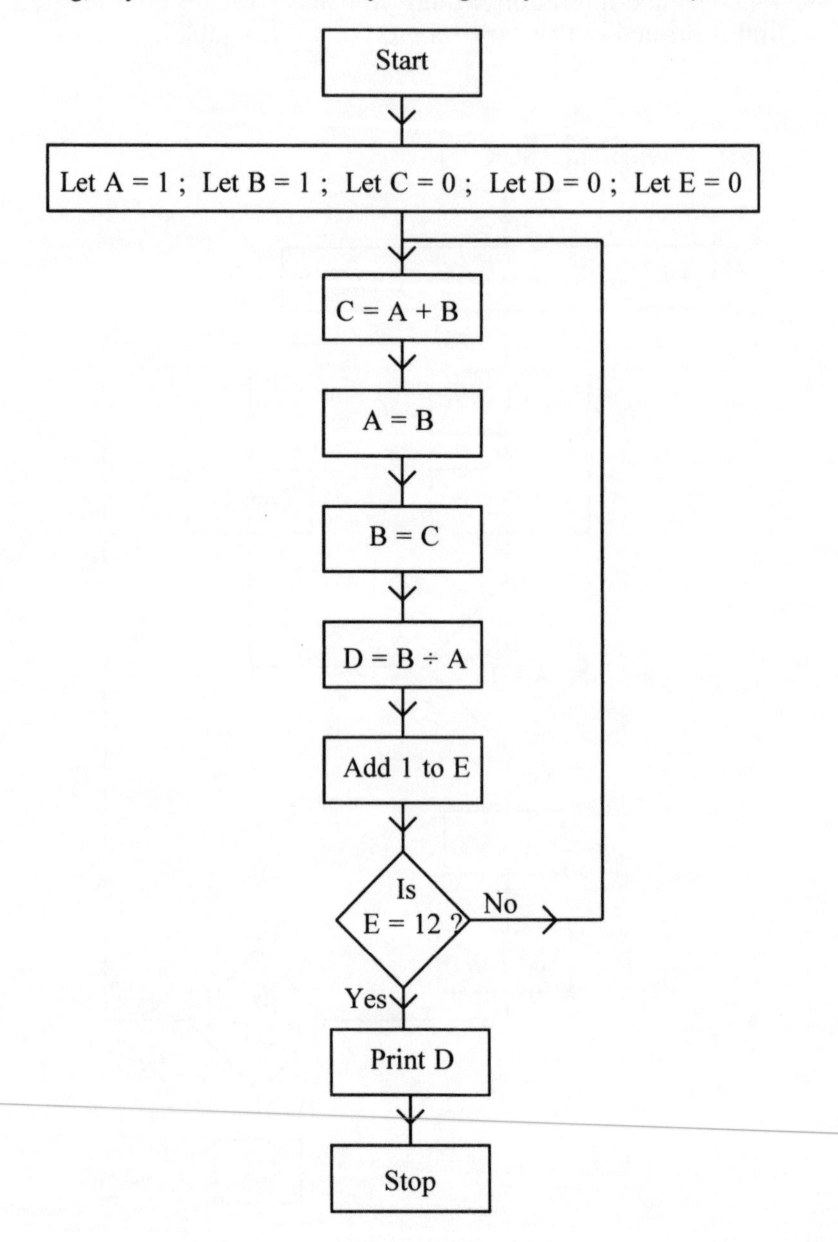

Figure 9

49. Use five columns for stores A, B, C, D and E to investigate the outcome of the flow diagram of figure 9. You will almost certainly need a calculator. Flow diagram experts could no doubt manage this flow diagram with fewer stores, but we are not experts, and this is difficult enough as it is.

50. If you added the instruction ‘Print A’ at a certain point in figure 9, it would print out the first 11 terms of the Fibonacci Sequence. After which instruction would you place ‘Print A’ in order to do so?

15

Topology

For this chapter you will need:

Pen and paper
Scissors and glue

Topology is the study of how things are connected. You will get a better idea of what this means as you work through the chapter, but you may remember the neat diagram of a battery, a bulb and a couple of switches in Chapter 11 on George Boole. The wires connecting these items were lovely and straight. In reality, it is most unlikely that they would have looked much like the diagram, but so long as they were connected in the way the diagram suggested, all would be well.

When writing this chapter, I thought that topology might be defined as the art of making simple things seem difficult; at times that will certainly appear to be so, but you may eventually realise that it makes you see things from a different standpoint.

One small piece of vocabulary: a 'simple closed curve' is a loop which does not cross over itself. For example, a C is not 'closed'; an 8 is not 'simple'; an O is the nearest my keyboard gets to one.

The chapter is divided into lots of smaller sections which do not necessarily get more difficult as you go through, so if you find one section too difficult, just go on to the next.

Lastly, for those of you who doodle a lot, you may be pleased to know that mathematicians have made a study of some of your doodles. It makes a change from psychologists doing so, I suppose.

Figure 1 (a) (b)

1. Figure 1(a) shows a doodle. Find where it starts and trace with your finger or a pen to where it finishes.

2. Do the same with figure 1(b).

A unicursal curve is a line which can be drawn with one flow of a pen or pencil, without taking the pen off the page and without going over a line twice. A point where the line starts, ends or crosses itself is called a node. If three lines meet at a point, it is called a 3-node; if 94 lines meet at a point it is called a 94-node.

3. Look at figure 1; can you find a 2-node?

4. What kind of nodes are the start and finish points of figure 1(a)?

5. What kind of nodes are the start and finish points of figure 1(b)?

6. What other kind of node is there in figure 1? (There is only one other kind here.)

7. Doodle a few unicursal curves and see if you can come up with a general rule as to where you find odd numbered nodes; i.e. 1-nodes, 3-nodes, 5-nodes etc.

8. Is it possible to draw a unicursal curve which has two 1-nodes, one 3-node and three 4-nodes?

In case you have not yet found the general rule, you will see that an odd numbered node indicates the starting or finishing point of a unicursal curve. While you are drawing, as you cross an existing line you create a node, but because you then leave that node, you create one which is even-numbered. The exception is when you start or finish, in which case you create an odd-numbered node.

9. Is the digit 8 a unicursal curve?

10. Where is the starting or finishing point of the digit 8?

Questions 9 and 10 are included to show you that you do not need an odd-numbered node to have a unicursal curve. In the case of the 8, it starts and finishes at the same point which could be any point along its length.

The ideas you have met so far can be used to simplify, and then solve, various problems. The classic example is the 'Bridges of Königsberg' problem.

The story goes that in Königsberg, there is an island where the River Pregel divides. There were bridges across the river as shown in figure 2. (There are now two more bridges.) At the time the problem was posed, it was the fashionable thing while out for your Sunday stroll to try to cross all the bridges without crossing any bridge twice.

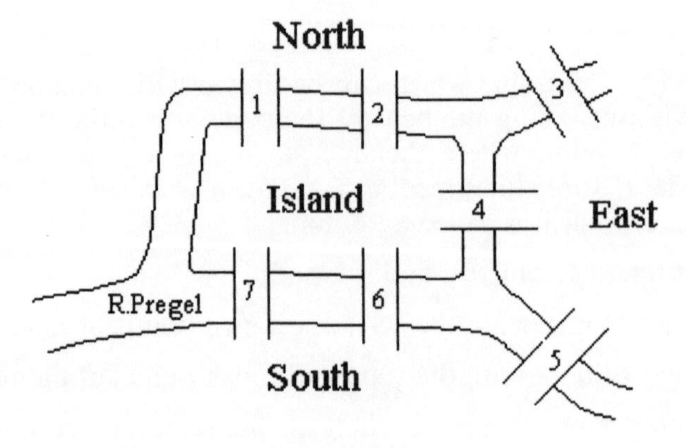

Figure 2

11. Would you be able to do it?

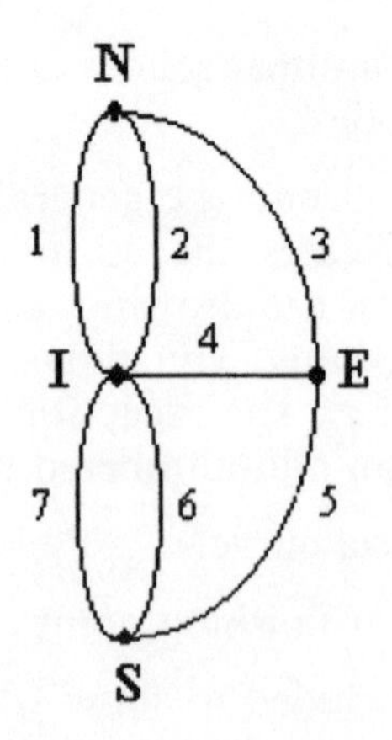

Figure 3

Figure 3 has changed the problem, but the connectivity of it has not changed. Imagine that there is a meeting point, or possibly a starting point, to the North, South and East of the island, and on the island itself. Imagine, too, that so many people go from meeting point to meeting point across the bridges that they wear paths to these points. The map of figure 2 would now look like figure 3 where N, S, E, and I are the North, South, East and Island meeting points, and paths 1 to 7 take you across the bridges numbered similarly.

12. Apply your knowledge of unicursal curves to figure 3 to see if the walk is possible.

Dominoes

A domino piece consists of a rectangle divided into two squares, each square being numbered. Imagine that in a small scale game, the numbers used are 0, 1, 2, 3 and 4. The problem is to see if you can arrange all the pieces so that any two squares from different pieces which touch each other have the same number. For example, if the piece with number 0 and number 1 on it is written as (0,1) then we can arrange pieces as: (0,1) (1,2) (2,4) (4,3)... etc. 1 touches 1; 2 touches 2 and so on.

The pieces for this set are:
(0,0) (0,1) (0,2) (0,3) (0,4) (1,1) (1,2) (1,3) (1,4) (2,2) (2,3) (2,4) (3,3) (3,4) (4,4).

(If you can remember back to chapter 7 on Pascal's Triangle, there are 15 pieces because the number of ways of choosing 2 from 5 is 10 (if each digit is used twice), and we also have 5 'doubles'.)

13. See if you can arrange these 15 domino pieces so that same numbers touch.

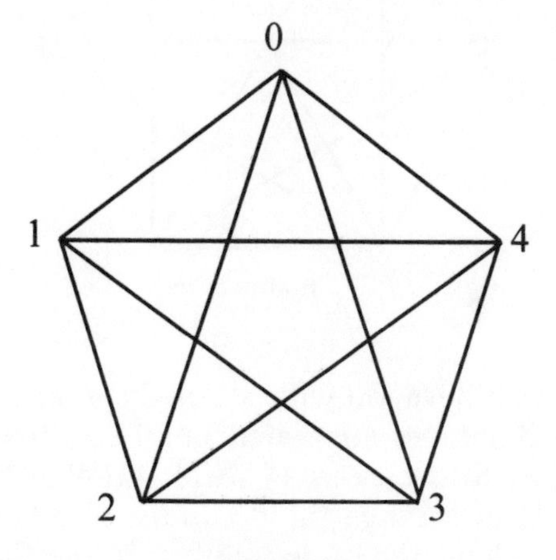

Figure 4

The unicursal curve of the pentagon and its diagonals, figure 4, will solve this problem for you, of course! Notice that the corners of the pentagon are labelled 0 to 4 and any side or diagonal of the pentagon represents a domino piece. For example, the side joining corner 1 to corner 2 represents the piece (1,2). Notice, too, that I have missed out the doubles, such as (0,0) and (1,1). This omission is put right in question 14.

Starting at 0, trace around the outside of the pentagon to 4; this represents the pattern: (0,1) (1,2) (2,3) (3,4).

14. In the order (0,1) (1,2) (2,3), where would you put the double 0, (0,0) and the double 1 (1,1) and the double 2 (2,2)?

Note that this problem is easily solved, so that putting the dominoes in order, we can ignore the doubles. You can put them in afterwards if you want to.

15. The order (0,1) (1,2) (2,3) (3,4) can be continued to solve the problem. Can you trace around the pentagon and its diagonals to see how this is done?

16. Is your solution to question 15 the only possible solution?

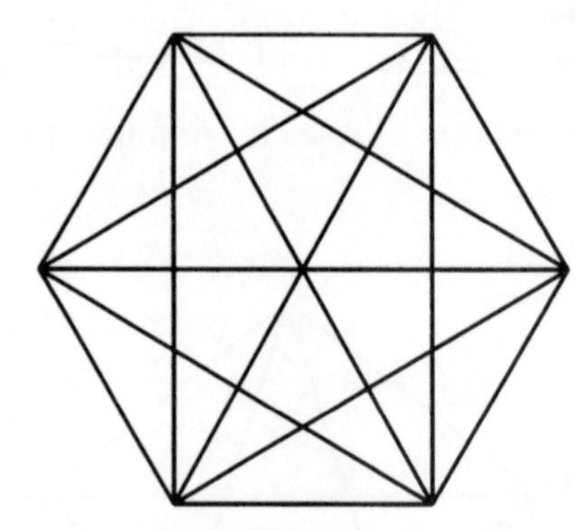

Figure 5

17. Figure 5 shows a hexagon with all its diagonals. You can use it to represent the dominoes numbered in all combinations from 0 to 5 in a similar way to questions 14 to 16. Look at figure 5 and decide if it is possible to arrange the dominoes numbered 0 to 5 so that touching pieces have the same number. (Note that this is the same as asking if figure 5 is a unicursal curve.)

18. It is not possible to arrange the pieces of question 17 in the necessary way, but it is possible to arrange the pieces numbered in combinations from 0 to 6; this is the common sort of dominoes you will find in the Pig and Whistle! Complete this question only if you have the time and patience to do so. Draw a heptagon – a 7-sided figure – which does not have to be drawn perfectly. Draw in all its diagonals and number its corners 0 to 6. Use the diagram to find out how to arrange this set of dominoes so that like numbers touch.

Remember that this method will not include the doubles at first. The pieces are:
(0,0) (0,1) (0,2) (0,3) (0,4) (0,5) (0,6) (1,1) (1,2) (1,3) (1,4) (1,5) (1,6) (2,2) (2,3) (2,4) (2,5) (2,6) (3,3) (3,4) (3,5) (3,6) (4,4) (4,5) (4,6) (5,5) (5,6) (6,6), 28 in all.

If you have such a set of dominoes, why not check your answer by experiment!

Mazes

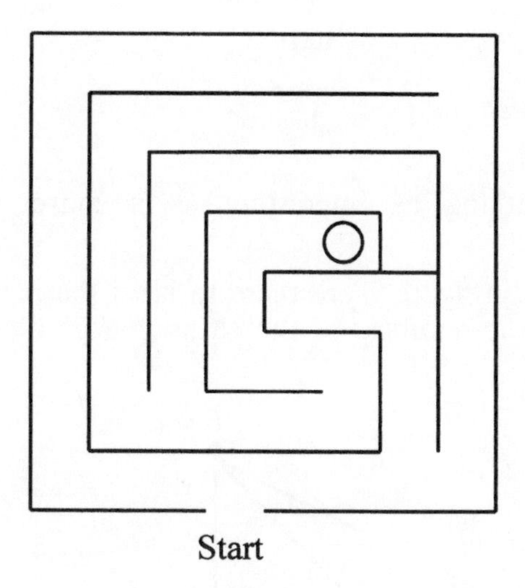

Figure 6

19. Follow from the start of the maze of figure 6 to the circle in the middle.

This is not a very difficult maze to follow; I found it much more difficult to design it than to follow it. The first step I took was to design the paths, not the walls.

20. Follow the paths of figure 7 and make sure that they agree with the same paths you would follow between the hedges or walls of figure 6. Ignore the numbers.

We now embark on something which mathematicians are good

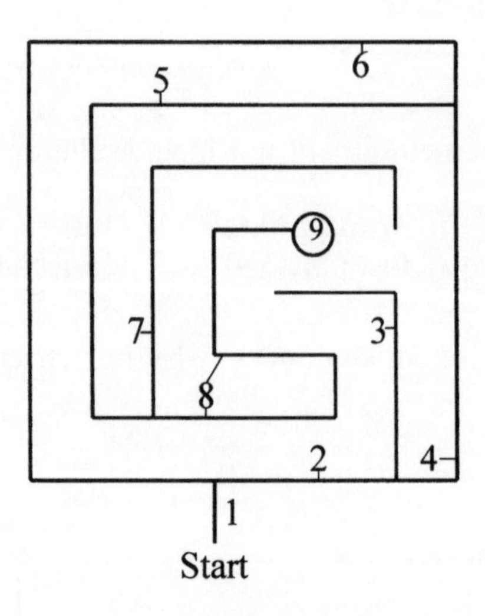

Figure 7

at: taking something as interesting as a maze and making it tedious!

21. The numbers in figure 7 are there to label the paths of the maze. Write down the numbers of the paths you go along to reach the centre.

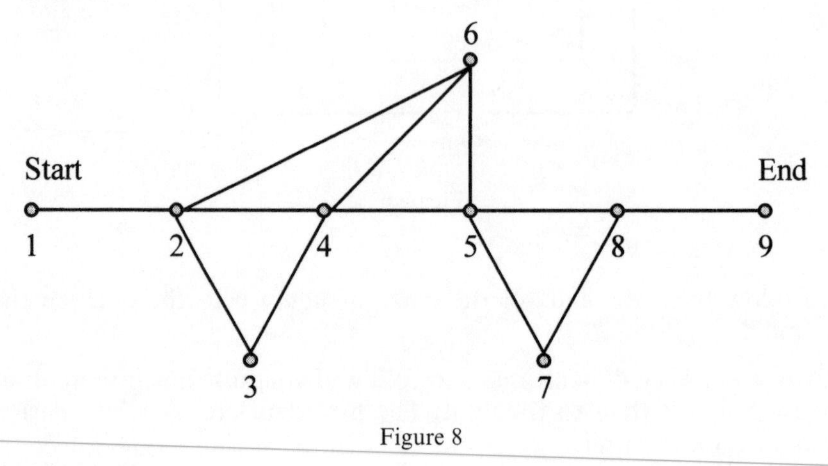

Figure 8

Figure 8 shows a trick of topology: the paths have been shrunk to points. Because of this you can show if it possible to get from one path to another. For example, you can get from path 2 to path 3 because they are joined by a line.

22. In figure 8, one of the lines has been missed off. Which one?

23. Examining figure 8 you will see that you may well go along some paths and get back to where you started; in other words, there are some loops. How many can you find, assuming that you have put in the missing line?

24. Now that you are clever enough to get to the middle of this maze very quickly, how many paths do you *not* go along?

25. Is it possible to get to the centre of this maze by walking along all the paths but not going along any path twice?

If you want to make up your own maze, here is a tip: draw a plan of the paths first then draw the walls between them (I used red ink for this) to give the shape of the maze.

Surfaces

If you have a piece of paper on which a simple closed curve has been drawn (that is a loop which does not cross over itself) (See figure 9) you cannot get from a point A inside the curve to a point B outside it without crossing the curve.

This would seem obvious until...

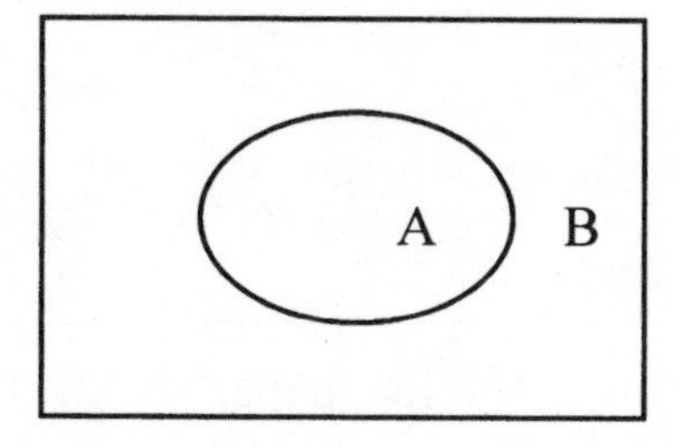

Figure 9

Figure 9 shows a point A inside a simple closed curve, and a point B outside it.

26. Cut out a strip of paper like figure 10 – it does not have to be exactly the same dimensions – and draw a line along its length. Glue its ends together so that A sticks to D, and B to C to form a belt or short cylinder.

27. The closed curve of figure 9 divides the piece of paper into two separate parts which we could call the inside and the outside. Does

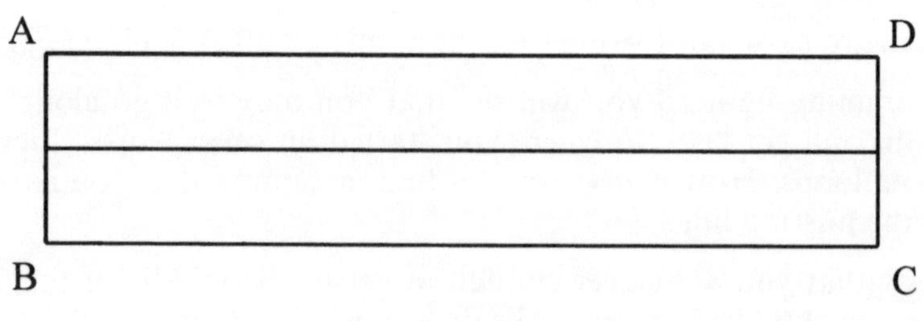

Figure 10

the line drawn on the belt you just made divide it into two separate parts?

28. Cut another strip of paper and draw a line along its length as you did for question 26, but this time when you glue the ends together, twist it so that A sticks to C and B sticks to D. The shape you now have is called a Möbius strip or Möbius band as in figure 11.

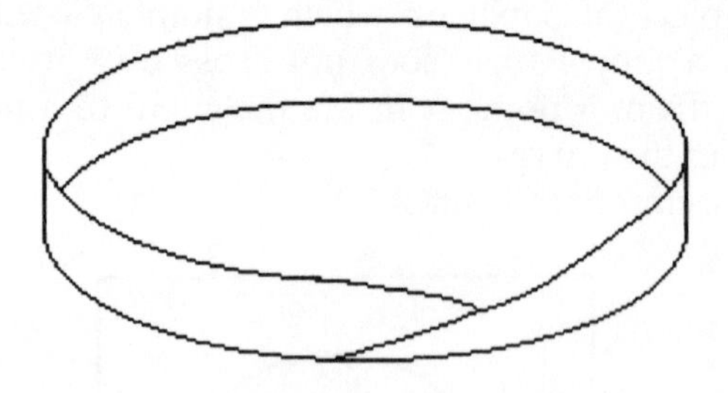
Figure 11

29. Does the line drawn on the paper divide the shape into two separate parts?

30. Put your finger on any point of the belt you made for question 26. Trace it around until you come back to where you started. Did you touch every part of the belt?

31. Do the same for the Möbius strip. Did you touch every part of it?

The Möbius strip is an example of a one-sided shape.

The cylinder has an inside and an outside. The fan belt of a car, and other drive belts, are this shape and wear out on the inside, the outside remaining much the same because it does not touch the machinery. I have seen drive belts powering farm machinery from old steam engines where the belts have been twisted into Möbius strips. The idea is that, having only one surface, the drive belt will wear all the way round and so last twice as long.

32. See if you can find a simple closed curve on your Möbius strip which does not divide it into two separate parts. If you cannot find one, go on to question 33.

33. Take a pencil or pen and continue the line you drew along the Möbius strip until you come back to the other end of that line. It should pass along the entire surface.

34. This line does not divide the Möbius strip into two separate parts. To prove it, take a pair of scissors and cut all the way along the line. The Möbius strip should not fall into two pieces, whereas the belt of question 26 would.

16

Numbers and Patterns

For this chapter you will need:

Pen and paper
A calculator, perhaps

If you asked someone what a pattern was, they would almost certainly tell you that it had something to do with wallpaper – or decoration of some sort. That is true, and in schools' mathematics, the first experience children have of pattern is in design.

However, pattern in numbers is also very important. It is really the initial step in algebra, since algebra – arithmetic algebra at least – is used as a means of expressing pattern.

In this chapter, the patterns are made up using dots, or crosses or whatever you choose; the number of the pattern is the number of dots used.

The most familiar group is the set of square numbers. 49 is a square number, not because of the way it is written but because we square 7 (7×7) to obtain it. The geometrical pattern comes when we arrange 49 dots in a square.

This study can be used to help children with their multiplication tables, but it can also help to familiarise us all with numbers. It is said that the great Indian mathematician Ramanujan saw numbers as his friends; he was familiar with them and their characteristics.

It is true that certain numbers do crop up frequently, for example 24 is in quite a few multiplication tables, but that is not the same as assigning magical powers to numbers. The number 13 is rather dull! Just an ordinary prime number. It is strange that it is associated with luck.

Much of the material for this chapter has come from the hours I have spent thinking of ways to reintroduce multiplication tables to children who have not yet learned them, but much, too, has come from David Wells' book *Curious and Interesting Numbers*.

Square, Rectangle and Other Numbers

```
*   *   *
*   *   *
```

Figure 1

6 dots can be arranged in a rectangle as in figure 1. This rectangle has no gaps or 'holes' in it.

1. Can you arrange 5 dots in a rectangle?

2. What shape can you arrange 4 dots in?

3. Draw dot patterns for each of these numbers and decide if they can be described as square numbers, rectangle numbers or other numbers:

(a) 3 (b) 7 (c) 8 (d) 9 (e) 11 (f) 12.

From question 3, you will see that 9 is a square number, 8 and 12 are rectangle numbers and that 3, 7 and 11 are others, which are called prime numbers. Numbers which are not prime – both square and rectangle numbers are such – are called composite numbers, numbers which are lumps of other numbers pushed together.

4. Allow space on your paper below and to the right of this question. Mark one dot. That is all!

In case you have not noticed, you have drawn a pattern of dots for the first square number, 1. Note that 1 is not a prime number.

5. Add more dots to your dot of question 4 to make a pattern for the next square number (which is 4).

6. How many dots did you add to your diagram when you did question 5?

7. Add more dots to your diagram of question 5 to make a pattern for the next square number (which is 9).

8. How many dots did you add to your diagram this time?

9. The next square number is 16; how many dots do you think you must add to your diagram so that it represents 16?

In case you are as good as me at missing the most obvious facts, the first square number is 1 ($= 1 \times 1$), the second is 4 ($= 2 \times 2$), the third is 9 ($= 3 \times 3$) etc.

10. You have probably realised that you can get from one square number to the next by adding an odd number. Which (and how many) odd numbers do you have to add to get 4, the second square number?

11. To get 4, the second square number, you have added two odd numbers, 1 and 3. How many odd numbers do you have to add to get 9, the third square number?

12. To get 9, you have added 1, 3 and 5, giving you the third square number. Continuing in this way, how many odd numbers would you have to add to get 400, the 20th square number?

13. To show your friends how clever you are, add together the first 10 odd numbers: (Do not add them, of course!)

$1 + 3 + 5 + 7 + 9 + 11 + 13 + 15 + 17 + 19.$

14. Remember that a prime number cannot be divided exactly by any number other than itself and 1. So 2 and 3 are the first two prime numbers, and higher than that, they usually come one more or one less than numbers in the 6 times table. Note that there is no hard or fast rule to this effect; it just happens that way. Here are the first 6 numbers in the 6 times table. Write down all the prime numbers which are 1 more or 1 less than these numbers: 6 12 18 24 30 36.

Triangle Numbers

The number 1 is taken to be the first triangle number, although one dot does not make much of a triangle. This way of looking at 1 helps more with the pattern of numbers than with the geometrical pattern.

```
*          *
* *        * *
           * * *
```

Figure 2

15. Figure 2 shows dot patterns for the second and third triangle numbers. What is the (a) second (b) third triangle number?

16. Remembering that 1 is the first triangle number, how many dots would you have to add to the first triangle number to get the second?

17. How many dots would you have to add to the second triangle number to get the third?

18. Use the pattern developed from questions 16 and 17 to calculate the fourth and fifth triangle numbers. Draw patterns of dots to check your answers.

Here are the first few triangle numbers so that you can check that you are not going too far astray: 1 3 6 10 15 21.

19. Add the first triangle number to the second.

20. Add the second triangle number to the third.

21. Add the third triangle number to the fourth.

22. What do you think the answer would be if you added the ninth triangle number to the tenth?

x o o o
x x o o
x x x o
x x x x

Figure 3

23. Figure 3 shows in diagrammatic form how the third triangle numbers, the o's, is added to the fourth, the x's, to make a square. Draw a similar diagram to show how the fourth and fifth triangle numbers add together to make a square.

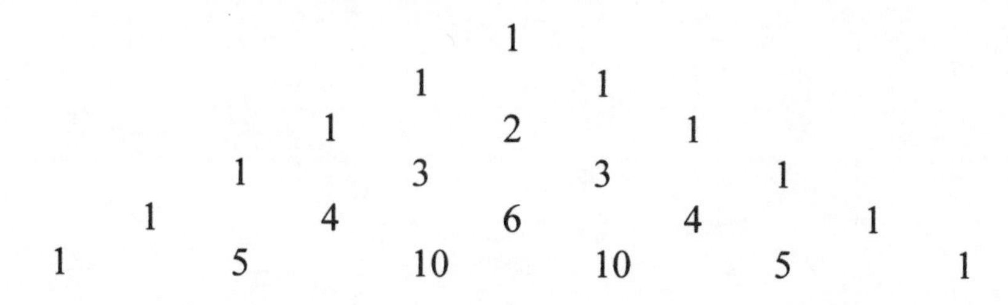

Figure 4

24. Figure 4 shows some of the first lines of Pascal's Triangle (see chapter 7). Stare at its diagonals until you can see the sequence of triangle numbers.

						1							=
				1	+	2	+	1					=
		1	+	2	+	3	+	2	+	1			=
1	+	2	+	3	+	4	+	3	+	2	+	1	=

Figure 5

25. Figure 5 shows another kind of number triangle which can obviously be continued. (It has nothing to do with Pascal or triangle numbers.) Add each row.

Note that you have seen figure 5 earlier, but in a disguised form. Here is how to strip it of its disguise. Look at the last line: add the last 1 to the first 2; add the last 2 to the first 3; add the last 3 to the 4. You will see that if you do this, the last line becomes:

$1 + 3 + 5 + 7$.

26. Add up these sums: (a) $1 + 2$; (b) $1 + 2 + 3$; (c) $1 + 2 + 3 + 4$.

27. This question hopes that you realise why question 26 was asked. The 100th triangle number is 5050. What is the sum of the first 100 counting numbers?

Pentagonal Numbers

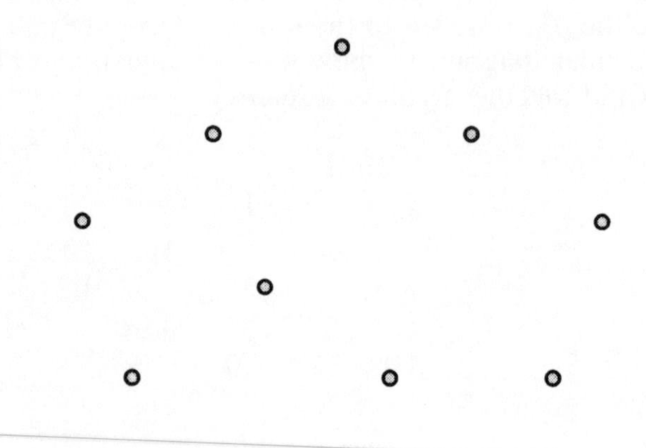

Figure 6

28. Besides being the first square and triangle number, 1 is also the first pentagonal number. Remembering that when you generate a new number in a sequence, you add to the existing one, how many dots do you need altogether for the second pentagonal number? (A pentagon has 5 sides.)

See figure 6 for help with question 28.

29. To obtain the third pentagonal number, another pentagon is added to the first two, with an extra dot along each side (see figure 6). What is the third pentagonal number?

30. To obtain the fourth pentagonal number, another pentagon is added to the other three of figure 6; it has three dots along each side. See if you can calculate the fourth pentagonal number.

To help you check your answers to questions 28, 29 and 30, and to save you further trouble, the first 7 pentagonal numbers are: 1 5 12 22 35 51 70.

31. The first 7 triangle numbers are: 1 3 6 10 15 21 28. There is a connection between the triangle numbers and the pentagonal numbers which is fairly simple, although it may take some time to work it out. See if you can spot the pattern with these hints:

The 4th pentagonal number is 22; the 4th triangle number is 10 and the 3rd is 6.
The 5th pentagonal number is 35; the 5th triangle number is 15 and the 4th is 10.
The 6th pentagonal number is 51; the 6th triangle number is 21 and the 5th is 15.

32. Calculate the 7th pentagonal number from two triangle numbers to see if you have the correct pattern.

33. The 8th triangle number is 36. What is the 8th pentagonal number?

Hexagonal Numbers

34. The first hexagonal number is 1. What is the second? (A hexagon has 6 sides.)

Wait a minute! This is getting too complicated. What we need is the method of differences.

35. Copy the first 6 square numbers, leaving a space between and below them. They are: 1 4 9 16 25 36.

36. What is the difference between the first two? That is, what is 4 – 1 ? Write your answer between and below the 1 and the 4.

37. What is the difference between the next two? (9 – 4 = ?). Write your answer between and below the 4 and the 9.

38. Continue the pattern, and use it to check that the next square number is 49.

39. Copy the first 6 triangle numbers then write down their differences as in questions 36, 37 and 38. They are: 1 3 6 10 15 21.

40. Use your answer to question 39 to continue the pattern one step further to check that the 7th triangle number is 28.

41. The first 6 pentagonal numbers are: 1 5 12 22 35 51. Use the methods of the last few questions to check that the 7th pentagonal number is 70.

42. The first 4 hexagonal numbers are: 1 6 15 28. What are the next two?

43. A hexagon has 6 sides and a triangle has 3, so it is not surprising that the hexagonal and triangle numbers are connected. Look at them and see if you can find the connection.

17

Triangles and Circles

For this chapter you will need:

Pen, pencil and paper
Compass (correctly 'compasses')
Ruler, protractor
Set square possibly

The main body of this chapter looks at some properties of triangles and a couple of circles associated with them. It has an Extra section at the beginning.

The Extra section at the beginning shows you how to draw a rhombus (or diamond) with a ruler and compass. In this sense, to a mathematician, a ruler is something you use to draw straight lines with, not something you use to measure lengths with. Once you can construct a rhombus, you can bisect angles, bisect lines at right angles, drop perpendiculars and all sorts of other wonderful things. This will enable you to do some of the later constructions using ruler and compass techniques.

However, do not despair if you do not want to do this section because all the constructions in the main part of the chapter can be done using various mathematical instruments, including a ruler for measuring. For example, you can bisect an angle – cut it in two – by a ruler and compass construction, or you can simply measure the angle with a protractor, divide the angle by two then mark it and draw it.

Ruler and Compass Construction Extra

Make an approximate copy of figure 1 to the left hand side of your paper.

Open your compass to a distance just less than AB or AC.

Throughout this construction, do not alter the setting of your compass.

Put your compass point on the point A, where the two lines meet.

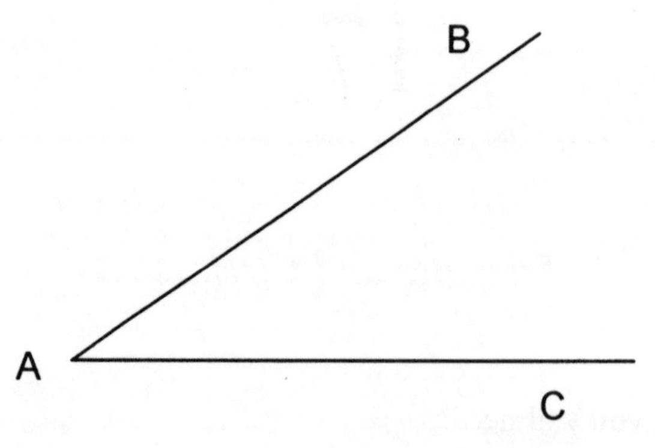

Figure 1

Make two marks with the pencil of your compass, one on the line AB near to B, and the other on the line AC near to C. From now on, take the first mark to be the point B and the second to be C.

Put your compass point on B and draw a wide arc below B, towards C and away from A.

Put your compass point on C and make a mark on the arc you have just drawn. Label this last mark as the point D.

If you were to – and you don't have to – join B to D and C to D, the shape ABDC would be a rhombus because all the lengths of its sides would be the same.

As a result, the line AD, if you drew it, would bisect the angle at A.

If you then joined B to C, the line BC would cut AD in half and at right angles. Also, the line AD would cut the line BC in half and at right angles.

To summarise: once you can draw a rhombus you can bisect an angle, draw lines at right angles to given lines and find the mid-point of a line.

Here is a variation on the last theme:

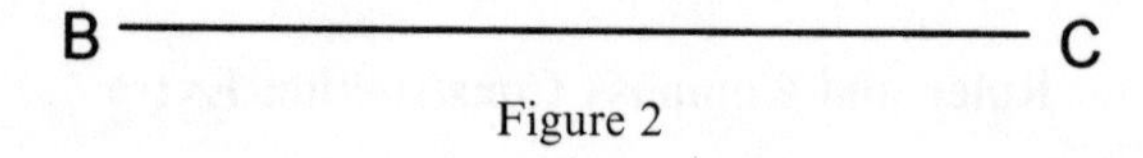

Figure 2

Draw a line as in figure 2, which is about 4 or 5cm long, so that there is about 5cm above and below it on your paper. Label the ends of this line as B and C.

Put your compass point on B and open it up to a distance roughly the length of BC. Do not change this setting throughout the construction.

Draw a wide arc above the line, and another below it.

Put your compass point on C and make two marks, one labelled A where your pencil cuts the upper arc, and another labelled D where your pencil cuts the lower arc.

Again you can see that if you were to join B to A to C to D to B you would have a rhombus, and that the two diagonals, BC and AD would cut one another in half and at right angles.

This is how to draw the perpendicular bisector of BC, the line which cuts BC in half and is at right angles to it. This line is sometimes called the mediator of BC.

The final ruler and compass construction will be left for you to work out, because it is very much like the last one. It is to 'drop' a perpendicular from a point to a line.

Start with a long line across the page, rather like BC in figure 2, but longer.

Mark a point where A was above the line.

Take A to be one corner of your rhombus and the line (like BC extended) to be one diagonal.

Use your compass to locate B and C and then D. The line AD will then start at A and meet the line (BC extended) at right angles.

This is called 'dropping a perpendicular from A to the line BC'.

Note: These constructions are easier to do than they are to describe! If you get into the way of doing them, practise a few of them yourself.

Not the Extra Section!

The constructions are all on triangles. You will get more out of the constructions if you use an acute-angled scalene triangle such as figure 3; this means a triangle with all its angles less than 90° and no two of its sides the same length. You may photocopy figure 3 if you wish, but the triangle should be bigger than this if possible.

Construction 1:

1. On an approximate copy of figure 3, make a mark halfway along each side of the triangle.
2. Draw a line from corner A to the mid-point of the side BC. Then

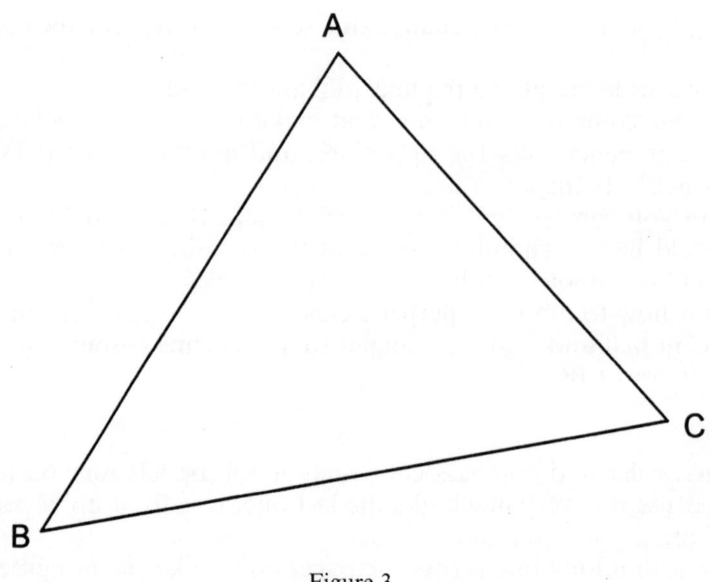

Figure 3

draw lines from B and from C to the mid-points of their opposite sides.

3. Each line you drew in question 2 is called a median, and the three medians should meet at a point; label this point G. G is a particular fraction of the way along each median. By measuring, see if you can find what this fraction is.

Note: If you cut a triangle from a piece of card, then draw the three medians across it, you should find that the triangle will balance on the edge of a ruler which is held under any of the medians. Theoretically, you should then be able to balance the triangle on the tip of a pencil placed under G; in reality, it is more likely to balance on the blunt end of a pencil.

G is called the centre of gravity, or centroid of the triangle.

Construction 2:

4. Draw another triangle like figure 3, as you did for construction 1.

5. Bisect each angle of this triangle. Do this by ruler and compass construction (three times) if you can. Otherwise, measure the size of each angle, divide the number of degrees by two then draw in the angle bisector – the line which cuts the angle in two.

6. The three angle bisectors should meet at a point. Label this point I. Put your compass point on I and draw a circle inside the triangle which just touches each side.

Notes: You will have done very well if your three angle bisectors do meet at a point, and particularly well if your circle does touch – not cross or miss – each side of the triangle. I is the in-centre of the triangle, and the circle is the in-circle.

Construction 3:

7. Start with another triangle as in figure 3, and this time draw the perpendicular bisector of each side of the triangle. Using a ruler and compass, this is the second construction of the first Extra section. If you want another way, do this: measure the length of each side so you can put a mark at the mid-point of each side. Then, using a set square or protractor (or credit card!), draw a line through each mid-point at right angles to the side towards the centre of the triangle.

8. Again, the three perpendicular bisectors should meet at a point. Label this point O. Put your compass point on O and draw a circle which passes through each corner of the triangle.

Notes: As with construction 2, you will have done very well if your three lines do meet at a point, and if the circle goes through – or near – the three corners of the triangle. The circle is called the circum-scribing circle and the point O the circum-centre.

Investigation: If you started Construction 3 with the scalene triangle of figure 3, the point O would be inside the triangle. If you would like to do a short investigation of your own into the location of O, start with an obtuse-angled triangle – one with one angle greater than 90° – or start with a right-angled triangle – one with one angle of 90°.

Extra Extra Section

Construction 4: (To cheat, have a look at figure 4!)

9. Start with the usual scalene triangle, but draw it as large as you can. Label its corners so that A is at the top, B is on the left and C is on the right.

10. Measure the size of angle A. You may want to turn the paper round to make one of the sides of the triangle horizontal; this makes it easier to use the protractor. Now divide the number of degrees by 3.
11. Trisect the angle A. This means draw two lines inside angle A so that they are equally spaced. The angle between each one will be the answer to question 10. Note that this cannot be done by ruler and compass construction.
12. Do the same for angles B and C.
13. Pick out and mark the point where the left-hand trisector of A meets the upper trisector of B.
14. Pick out and mark the point where the lower trisector of B meets the lower trisector of C.
15. Pick out and mark the point where the upper trisector of C meets the right hand trisector of A.

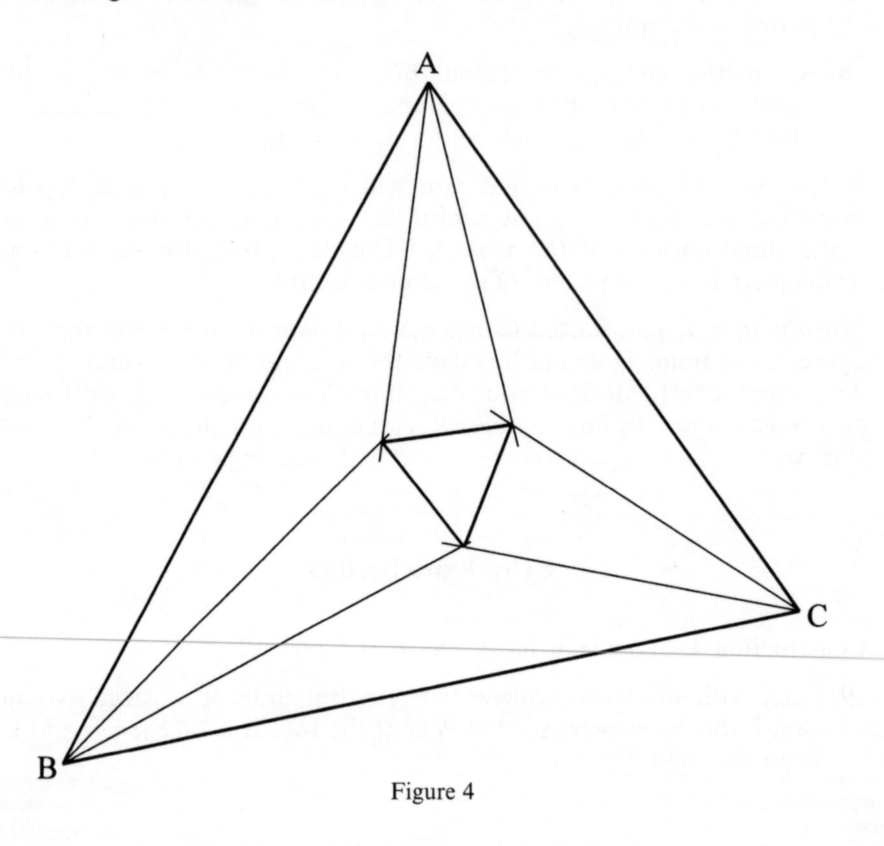

Figure 4

16. Join the three points which you marked in questions 13, 14 and 15 to make a triangle. What kind of triangle is this?

Note: I find it an amazing fact that you can take any triangle, trisect each angle, and points where the lines of trisection meet will always form an equilateral triangle no matter what shape of triangle you started with. It was discovered in 1899 by one F. Morley.

18

Base Two

For this chapter you will need:

Pen and paper

This chapter is concerned only with base 2, and of course base 10, which is the number base we all use. Some of the arithmetic may get a little difficult, but many of the answers are given to you in the text, and they are always available at the end of the book, too.

I have included a couple of examples where base 2 numbers are used, but I have, as usual, left out many more. I have not put too much stress on computers because, although they use base 2 for all their operations, the ordinary user like you and me does not have to be aware of this. We feed a computer with base 10 numbers, and it translates them into base 2, fiddles around with them, then gives them back to us in base 10 when it has finished with them.

I have been a little cavalier in my use of words throughout this chapter, so perhaps I should deal with its vocabulary here. Base 10, which is our normal system, is sometimes called the decimal system (from the Latin for 10 which is *decem*; remember that *Decem*ber was the tenth month before someone decided to put a couple more in). Base 2 is also called the binary system (from the Latin *bini* meaning 'in twos'). I have intermingled the terms 'base 2' and 'binary' so you must be prepared for either.

If you become confused by base 2 numbers, go ahead to question 16, which is intended to summarise the work, but may help you to understand it.

In base 10, our normal counting system, we use the digits 0, 1, 2, 3, 4, 5, 6, 7, 8 and 9. Imagine you are counting; when you get to 9 and run out of digits, you write 10. This means 1 ten and 0 units; it is, of course, a count of all your fingers and thumbs. If you keep counting, when you get to 99 and run out of digits again, you write 100, which means 1 hundred, 0 tens and 0 units. ($100 = 10 \times 10 = 10^2$). If, like

Anton Bruckner, you keep on and on counting, when you next run out of digits at 999, you simply create another place and write 1000, 1 thousand, 0 hundreds, 0 tens and 0 units.
($1000 = 10 \times 10 \times 10 = 10^3$).

Computers, deep inside their works at least, use only two digits, 1 and 0. Two digits force them to work in base 2. The base 2 counting system is much like our base 10 system except that you run out of digits much quicker.

1. Here we start counting in binary (base 2). Notice what happens when we run out of digits. Copy the list, which has been started so that you get an idea of what is happening, then continue it for at least another 8 numbers if you can.

0 1 10 11 100 101 110 111 1000 . . .

2. In the decimal system, we can multiply by 10 by moving the digits along one place, so $13 \times 10 = 130$. In base 2, what is $1\ 0\ 1\ 1 \times 1\ 0$?

3. The answer to question 2 is 1 0 1 1 0 ; how many times bigger is 1 0 1 1 0 than 1 0 1 1?

4. In base 10, $13 \div 10 = 1.3$; what is the .3 of 1.3?

5. In base 2, $1\ 0\ 1\ 1 \div 1\ 0 = 1\ 0\ 1\ .\ 1$. What fraction is the .1 of 1 0 1 .1?

The answer to question 5 is ½.

6. What is 1 0 1 . 1 take away one half?

Notes: In base 10, multiplication and division by 10 are easy. In base 2, multiplication and division by 2 are easy. In base 10, the first decimal place is tenths. In base 2 it is the first bicimal place which is ½.

Here is an old chestnut of a problem which I will later relate to base 2, thus destroying any interest you may have in it!

Mr and Mrs Feed lived on a farm. Mrs Feed looked after the hens and ducks with so much enthusiasm, she was known locally as Mrs Chicken Feed.

One day, Mrs Chicken Feed set off for market with eggs in her basket. She called on her friend Mrs Prima, who bought half her eggs and half an egg. She then went on to her friend Mrs Doss, who bought half the eggs she had left, and half an egg. Her next call was on her friend Mrs Dry who bought half the eggs she had left and half an egg.

When Mrs Chicken Feed got to market, she met her friend Mrs Agora, who bought half the eggs she had left, and half an egg. By the time she parted from Mrs Agora, Mrs Chicken Feed had only one egg left, so she went home to have it for her tea, pleased that she had sold all but one of her eggs, and that she had not broken one single egg all day.

7. How many eggs did Mrs Chicken Feed start with? (If you do not know the answer, read on!)

Here is the story, and the answer, in base 2 as well as in base 10:

Mrs Chicken Feed started with 31 eggs. In base 2 this is 1 1 1 1 1 eggs.

8. Using base 2 division (see question 5), what is half of 1 1 1 1 1?

9. Mrs Prima took half the eggs and half an egg. Like Mrs Prima, take half an egg away from the answer to question 8.

When Mrs Chicken Feed arrived at Mrs Doss's, she had 15 eggs, which in base 2 is 1 1 1 1 eggs (the answer to question 9).

10. Using base 2 division, what is half of 1 1 1 1?

11. Mrs Doss took half the eggs and half an egg. Take half an egg away from your answer to question 10.

When Mrs Chicken Feed arrived at Mrs Dry's, she had 7 eggs, which in base 2 is 1 1 1 eggs (the answer to question 11).

12. Using base 2 division, what is half of 1 1 1?

13. Mrs Dry took half the eggs and half an egg. Take half an egg away from your answer to question 12.

When Mrs Chicken Feed arrived at market, she had 3 eggs, which is 1 1 in base 2 (the answer to question 13).

14. Using base 2 division, what is half of 1 1?

15. Mrs Agora bought half the eggs and half an egg. Take half away from your answer to question 14.

You should find that Mrs Chicken Feed ended up with 1 egg, either in base 10 or in base 2.

Since the story of Mrs Chicken Feed is an old one, it is likely that anyone at the market who weighed anything did so on balancing scales using weights in pounds and ounces. 'Pounds' was (is) abbreviated to 'lb' (from the Latin *librum* meaning a pound; this also gives us the £, which is an L, for pounds-money). And 'ounces' was (is) abbreviated to ‘oz’, presumably because ‘oz’ is easier to write than ‘ounces’.

There were (are) 16 ounces to 1 pound and the weights which a market trader would have used were 1oz, 2oz, 4oz, 8oz. If a merchant had to weigh heavier amounts, he/she would also have had 1lb, 2lb, 4lb weights.

Weight (oz)	**8oz**	**4oz**	**2oz**	**1oz**
1				1
2			1	0
3			1	1
4				
5		1	0	1
6				
7		1	1	1
8				
9				
10				
11				
12				
13				
14				
15				

Table 1

16. Table 1 shows how you made up any weight in whole ounces between 1oz and 15oz using the weights given. Copy and complete it.

Note that this table enables you to translate base 10 numbers into base 2 numbers, and vice versa. Note, too, that if you take 1lb to be 16oz, and 2lb to be 32oz and so on, you can continue to translate larger numbers. For example, the base 10 number 31 can be made up from 16 + 15. 16 is 1 0 0 0 0 in base 2, and 15 is 1 1 1 1; adding them together, 31 becomes 1 1 1 1 1 in base 2.

Base Two Extra

Figure 1 shows a short piece of ticker tape:

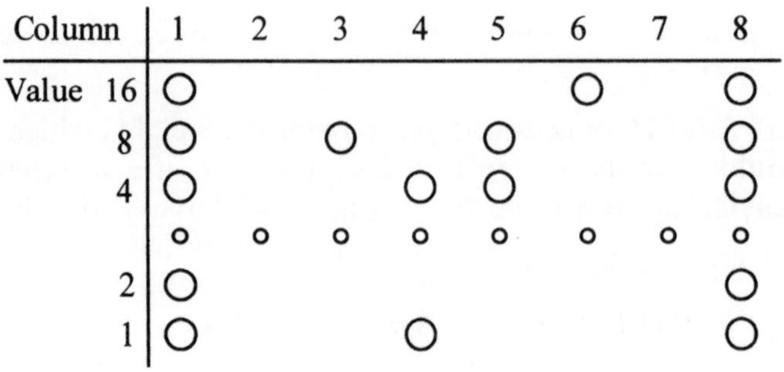

Column	1	2	3	4	5	6	7	8
Value 16	O					O		O
8	O		O		O			O
4	O			O	O			O
	o	o	o	o	o	o	o	o
2	O							O
1	O			O				O

Figure 1

If you ran this tape through a tape reader, you would find that it had a simple 4-letter message. In real life it would consist of a paper tape in which many holes have been punched.

The left-hand column of numbers would not normally be there, but I have put it in to help you translate the tape into numbers, which you can then translate into letters.

The row of small holes between 4 and 2 engage a cog which drives the tape through the tape reader; they are not part of the message!

Column 1 consists of a complete row of holes; this indicates the start of the message.

Column 2 is blank, which indicates a space.

Column 3 has a hole in the '8' row, so it represents the number 1 0 0 0, which is 8 in base 10.

Column 4 has holes in the '4' and '1' rows, so it represents the number 1 0 1, which is 4 + 1 = 5 in base 10.

Column 5 has holes in the '8' and '4' rows, so it represents 1 1 0 0, which is the number 8 + 4 = 12 in base 10.

Column 6 has a hole in the '16' row, so it represents 1 0 0 0 0, which is 16 in base 10.

Column 7 is blank, so it indicates a space.

Column 8 consists of a complete row of holes; this indicates the end of the message.

We now have to translate the numbers into letters; this is done in base 10 by this cipher:

A	B	C	D	E	F	G	H	I	J	K	L	M
1	2	3	4	5	6	7	8	9	10	11	12	13

N	O	P	Q	R	S	T	U	V	W	X	Y	Z
14	15	16	17	18	19	20	21	22	23	24	25	26

The numbers to translate are: 8 5 12 16 which are H E L P

Figure 2 is a picture of a long piece of tape for you to translate. I have been less than helpful in not numbering the rows, so you may have to write the numbers first in base 2, then use the table of weights to translate into base 10, then the cipher to translate into letters.

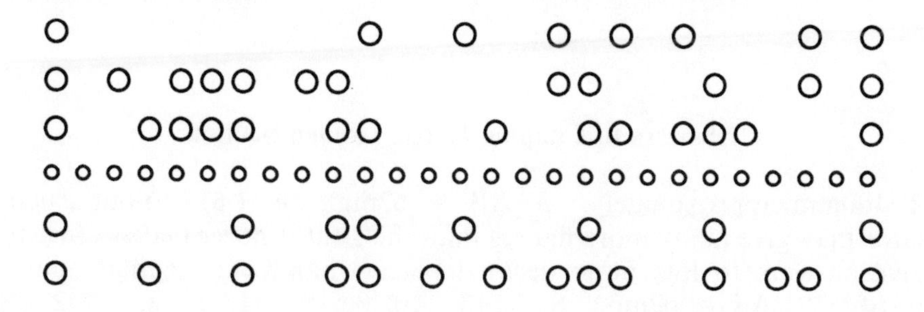

Figure 2

Books referred to in the preparation of this book were:

Gardner, M., *Mathematical Circus* (Penguin Books, 1968)
Land, F., *The Language of Mathematics* (John Murray, 1960)
Lietzman, W., *Visual Topology* (Chatto & Windus, 1965)
Somerville, D., *Analytical Conics* (Bell & Sons, 1956)
Stewart, I., *The Magical Maze* (Weidenfeld & Nicolson, 1997)
Wells, D., *Curious and Interesting Numbers* (Penguin Books, 1986)

ANSWERS

Answers to Chapter 1, The Golden Section

2. 100mm approximately. **3**. AB = 62mm. **4**. 1.613. Your calculator may give many more figures than this, but it never makes sense to give answers to lots more decimal places than your original figures justify. **7**. AT = 62mm. **8**. 1.613. **10**. 0.618. **11**. 0.618. **12** 2.618 **13**. 2.618. **14**. ST = 24mm (approx). **15**. 39mm. **16**. 63mm. Note that to get question 16 to be 62mm, the length of AB, you would have to measure ST to be 23.6mm. This degree of accuracy is obviously not possible. **17**. 9mm – almost impossible to guarantee accuracy here. **18**. AB = 65mm. **19**. BC = 40mm. **20**. 1.625. **21**. BD = 40mm. **22**. AD = 25mm. **23**. 1.6 **24–31**. See Figure 1.

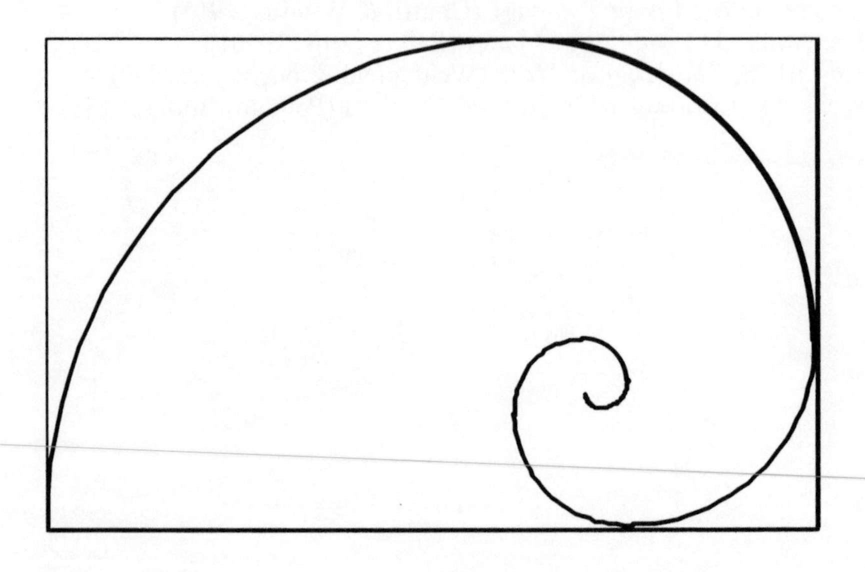

Figure 1

Answers to Chapter 2, The Fibonacci Sequence

1. 1, 1, 2, 3, 5, 8, 13, 21, 34, 55, 89, 144 ... etc. Each term is the sum of the previous two. **2**. 1, 2, 1.5, 1.666..., 1.6, 1.625, 1.615, 1.619. 1.618 etc. **3**. Even. **4**. Even. **5**. Odd. **6**. Every third number is even. **7**. The first two terms are odd, so O + O = E; the third term is even. The fourth term is O + E = O. The fifth term is E + O = O. The sixth term is O + O = E and we are back where we started. **8**. Every fourth term is divisible by 3. **9**. Every fifth term is divisible by 5. **10**. Every sixth term is divisible by 8. **11**. The numbers which divide into the terms increase in the Fibonacci Sequence. **12**. e.g. If n = 10, G^{10} = 122.966; 122.966 ÷ 2.236 = 54.993... which is 55 to the nearest whole number. **13**. e.g. $T_2 \times T_4 = 1 \times 3 = 3$; $T_3 \times T_3 = 2 \times 2 = 4$; $T_2 \times T_4 - T_3 \times T_3 = 3 - 4 = -1$.

Answers to Chapter 3, Exponential Growth

1. 220, 233, 247, 262, 277, 294, 311, 330, 349, 370, 392, 415, 440. **2**. The range of one octave is 440 – 220 = 220Hz. E at 330Hz is halfway up the scale (330 = 220 + 110). **3**. 7 semitones. **4**. D. **5**. A to D is 294 – 220 = 74Hz. 74 out of 220 is one third – within our limits. **6**. 77.5°. **7**. 32. **8**. 16, 8, 4, 2, winner! **9**. ½ or 0.5. **10**. £108. **11**. 1.08. **12**. £116.64. **13**. £215.89.

Answers to Chapter 4, Fractions.

1. 0.333... and 0.6666... **2**. 0.142857...; 0.285714...; 0.428571...; 0.571428...; 0.714285...; 0.857142... The same 6 digits appear in the same order, starting at a different place. **3**. 0.1111...; 0.2222...; 0.3333...; 0.4444...; 0.5555...; 0.6666...; 0.7777...; 0.8888... **4**. $\frac{2}{3} = \frac{6}{9}$ **5**. 0.0909...; 0.1818...; 0.2727...; 0.363636...; 0.4545...; 0.545454...; 0.6363...; 0.7272...; 0.8181...; 0.9090... **6**. (a) $\frac{1}{15}$ = 0.0666...; (b) $\frac{2}{15}$ = 0.1333...; (c) $\frac{3}{15} = \frac{1}{5}$ = 0.2 or 0.19999... **7**. $\frac{1}{17}$ = 0.0 588 235 294 117 647... **8**. There are 6 different remainders when dividing by 7, excluding 0, which denotes exact division; there are 16 remainders when dividing by 17. **9**. $\frac{3}{11}$; $\frac{1}{3}$; $\frac{9}{11}$; $\frac{2}{3}$. **10**. 0.2727...; 0.3333...; 0.8181...; 0.6666... **11**. $\frac{1}{7}$ because the fraction given is the sequence of digits from $\frac{1}{7}$ divided by 999 999. **12**. 0.123123... **13**. 0.8282... **14**. 0.62816281... **15**. 142857; 285714; 428571; 571428; 714285; 857142. **16**. 1 176 470 588 235 994; this is the same cyclic sequence of digits, starting with the digit which gives the next highest number. **17**. $\frac{3}{17}$ The numerator of the fraction is the same cyclic sequence of digits, starting with the next highest number after question 16. **18**. 999. **19**. 99 999 999. **20**. 999 999 999. **21**. 27 **22**. 72. **23**. 81. **24**. If each of the numbers in questions 21, 22 and 23 are the numerators of fractions with 9999... etc. as denominators, they will cancel to a simpler fraction. If they can do this, they must be divisible by 9, because the denominator is. Any number which is divisible by 9 has the sum of its digits also divisible by 9. **26**. The first and last columns make it clear that the digits 1 to 8 appear twice, and 0 and 9 appear once in each column. These digits sum to 81.

Answers to Chapter 5, Two-Dimensional Shapes

1. (3,4) **2–6**. See figure 2. **7**. The parallelogram has no lines of

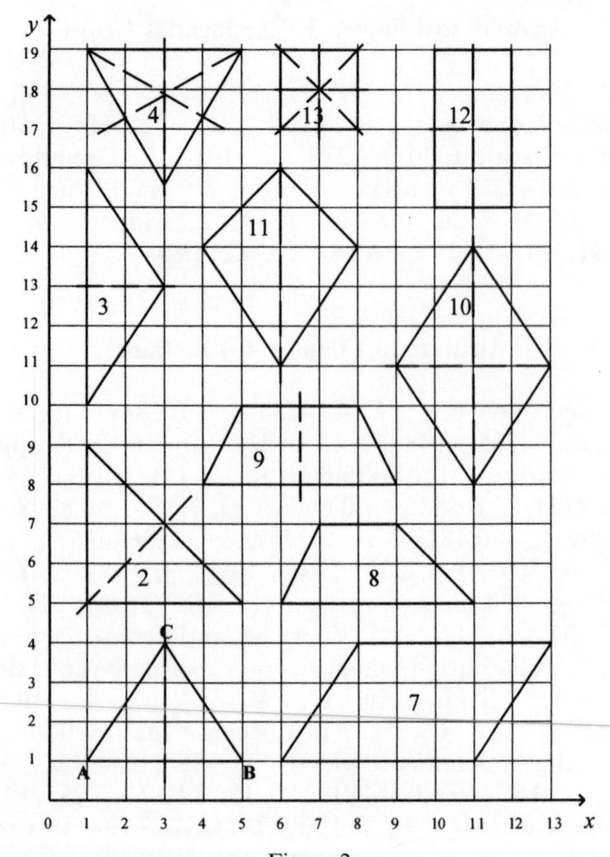

Figure 2

symmetry but has rotational symmetry of order 2 about the point where the diagonals cross. **8**. No symmetry at all. **9**. A line of symmetry bisecting the two parallel lines. **10**. The two diagonals are lines of symmetry and there is rotational symmetry of order 2 about the point where they cross. **11**. A line of symmetry through the top and bottom corners. **12**. Two lines of symmetry, one along the line where $x = 11$ and the other where $y = 17$. There is rotational symmetry of order 2 about the point where the lines of symmetry cross. Note that the diagonals are not lines of symmetry. **13**. Four lines of symmetry: the lines where $x = 7$ and where $y = 18$ and the two diagonals. There is rotational symmetry of order 4 about the point where the lines of symmetry cross. **14**. Yes! Arrange the quadrilaterals so that four different angles fit around a point. Because the angles are about a point, and the sum of the angles of a quadrilateral add up to 360°, there will be no gaps.

Answers to Chapter 6, Three-Dimensional Shapes

1. Rotational symmetry of a three-dimensional shape is about an axis. **2**. Yes, through any pair of opposite corners. **3**. Solid shapes reflect in a plane, as we do in a mirror, which is a plane. **4**. Yes; any plane which can contain four corners of the octahedron. Also any plane which contains two corners on opposite extremes of the octahedron and passes through the mid-points of two opposite edges. **6**. Consider your own reflection in a mirror – plane symmetry: when you raise your right hand, your image raises its left hand. If a mirror were like a slide projector – point symmetry – your image would appear upside down. **7**. The symmetries of a cube are much the same as those of an octahedron. **8**. 8 – that is what the word 'octahedron' means! **9**. 6. **10**. 12. **11**. $8 + 6 - 12 = 2$. **12**. 6. **13**. 8. **14**. 12. **15**. $6 + 8 - 12 = 2$. **17**. 12. **18**. 20. **19**. 20. **20**. 12. **21**. $12 + 20 - ? = 2$; $? = 30$. **22**. 30, the same as its dual. **24**. Face: 4; Corners: 4; Edges: 6; Euler: $4 + 4 - 6 = 2$. **25**. The tetrahedron is its own dual. **26**. Yes. **27**. When you place any of the shapes on a surface so that one corner only is touching that surface, the opposite corner should always be the same height above the surface. This height is the diameter of the sphere which can surround the Platonic Solid.

Answers to Chapter 7, Pascal's Triangle

3. Row 0: 1; Row 1: 2; Row 2: 4; Row 3: 8; Row 4: 16; Row 5: 32; Row 6: 64; **4**. The total for row x is 2 multiplied by itself x times. **5**. 2^5 **6**. The total is doubled because all numbers except 1 in each row are added twice, then two more 1s are added. **7**. 1; (1+2=) 3; (3+3=) 6; (6+4=) 10; etc. The difference between successive numbers increases by 1 each time. **8**. 5. **9**. 10. **10**. 1. **11**. 1. **12**. 5. **13**. 10. **14**. In how many ways can you select 3 things? – 4 things? – 5 things? **15**. Row 5 of Pascal's Triangle: 1, 5, 10, 10, 5, 1 gives the answers to questions 11, 12, 13 and 14. **16**. 4. **17**. 5. **18**. (a) 2; (b) 1; (c) 1. **19**. (a) 1; (b) 1; (c) 2 – HT and TH. **20**. HHH (1 way for 3H); HHT, HTH, THH (3 ways for 2H and 1T); HTT, THT, TTH (3 ways for 1H and 2T); TTT (1 way for 3T). **21**. 21, the 3rd number in row 7 of Pascal's Triangle. **22**. No. **23**. 20. **24**. 12. **25**. ½. **26**. Yes.

Answers to Chapter 8, Probability

2. (a) 4; (b) 52; (c) $\frac{4}{52} = \frac{1}{13}$. **3**. $\frac{13}{52} = \frac{1}{4}$. **4**. $\frac{1}{52}$. **5**. $\frac{1}{6}$. **6**. $\frac{48}{52} = \frac{12}{13}$. **7**. $\frac{39}{52} = \frac{3}{4}$. **8**. $\frac{51}{52}$. **9**. $\frac{5}{6}$. **10**. 1. **11**. 0. **12**. $\frac{4}{52} = \frac{1}{13}$. **13**. ? **14**. (a) 4; (b) 52; (c) $\frac{4}{52} = \frac{1}{13}$. **15**. (a) 3; (b) 51; (c) $\frac{3}{51} = \frac{1}{17}$. **16**. (a) 4; (b) 51. (c) $\frac{4}{51}$. **17**. $\frac{1}{6}$. **18**. $\frac{5}{6}$. **19**. $\frac{1}{6}$. **20**. A 6 with the first throw and not a 6 with the second throw. **21**. $\frac{1}{6} \times \frac{5}{6} = \frac{5}{36}$. **22**. Not a 6 with the first throw and a 6 with the second. **23**. $\frac{5}{36}$. **24**. Not getting a 6 with either throw. **25**. $\frac{5}{6} \times \frac{5}{6} = \frac{25}{36}$. **26**. 1. **27**. It is certain that all outcomes will happen. P(Certainty) = 1. **28**. 6. **29**. 1. **30**. 1. **31**. 15; the 3rd term in row 6 of Pascal's Triangle. **32**. 1. **33**. $\frac{1}{15}$.

Answers to Chapter 9, Statistics

2. For the graph of the letter frequency, see figure 3. **7**. YOU MUST BE VERY CLEVER IF YOU CAN DECODE THIS MESSAGE.

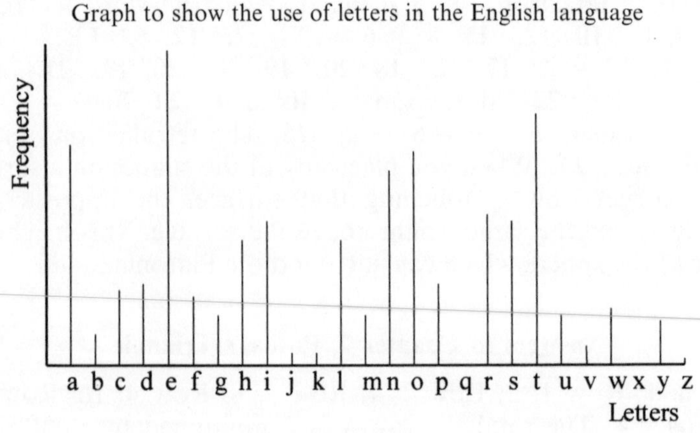

Figure 3

8. qyn a juwz a bm vyoq kryvyo.
The final coded message is: The English army had to find a way of breaking the codes, so they asked a man called Alan Turing to do it for them. He worked with lots of scientists and mathematicians and in six months they had made a machine which could break the enemy code. It was the first computer in the world and was called the Colossus. It took up most of a large room but was not nearly as powerful as the desktop computers we use today.

Answers to Chapter 10, Conic Sections

3. The sum of the distances from each focus to any point on the ellipse is constant. **4**. 100mm. The constant distance of question 3 is 100mm, the major axis of the ellipse. **5**. The closer the foci, the nearer the ellipse is to a circle. **6**. The further apart the foci, the flatter the ellipse. **8**. The beam of light from a spotlight is in the shape of a cone. Cutting the cone by a plane – the stage – gives an elliptical shape. **10**. Question 9 uses the result of question 3. **11–17**. Your completed figure should look something like figure 4. The curve is shown by a solid line; the tangents to the curve are shown by the

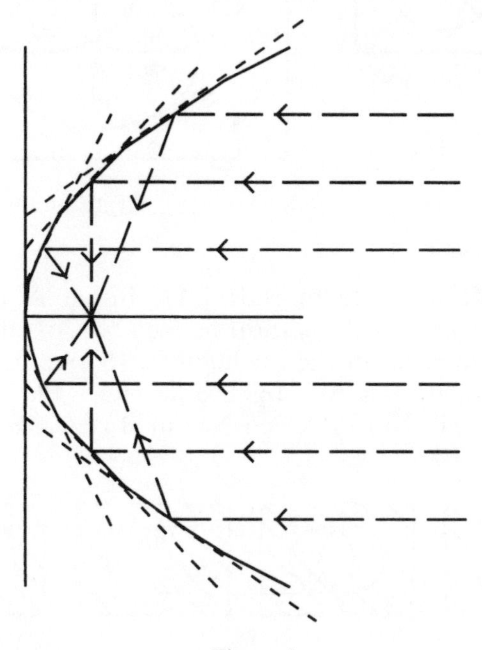

Figure 4

dotted lines; the rays of light – or whatever – are shown by the dashed line. The focus of the parabola is at the point (1,0). **19**. The path of whatever you have thrown is much the same as the curve of figure 4 shown by the solid line, but turned through 90° clockwise.

Answers to Chapter 11, George Boole

1. (a) 10 + 15 = 25 if no one attends both courses; (b) 15 if all members of the French course attend maths; (c) 5 members at least (25 – 20 = 5) would have to attend both courses. **2**. Members of Ferndown U3A. **3**. l, m, n, x, y, z. **4**. p, q, r, x, y, z. **5**. They attend both French and maths. **6**. They attend either French or maths. **7**. Members of Ferndown U3A who do not attend maths or French. **8**. See fig 5(a); (a) 5 grow both; (b) 12 grow just flowers; (c) 8 grow just vegetables. **9**. See fig 5(b); 3 do not use any of these

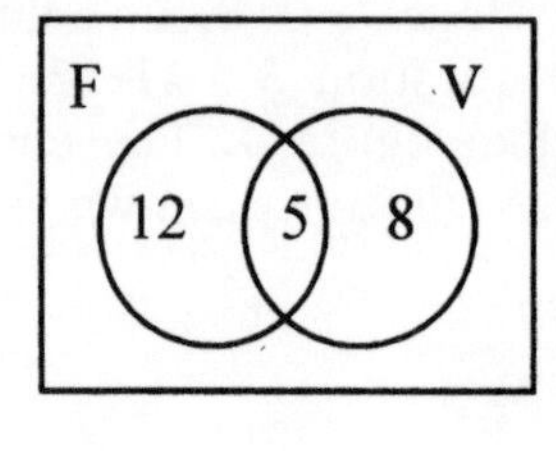

Figure 5(a)

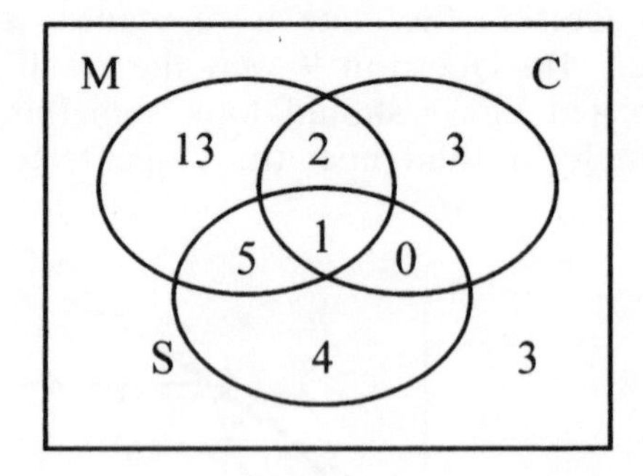

Figure 5(b)

cameras. **10**. All men are mortals. **11**. Inside A but outside B. **12**. Dogs? Any*thing* which is not a member of the cat family. **13**. Tigers? Any member of the cat family. **15**. See fig 6(a). **16**. See fig 6(b). **17**. Not in A is A′. **18**. See fig 6(c). **19**. Not in B is B′. **20**. See fig 6(d). **21**. Not in A and Not in B is $A' \cap B'$. **22**. It was

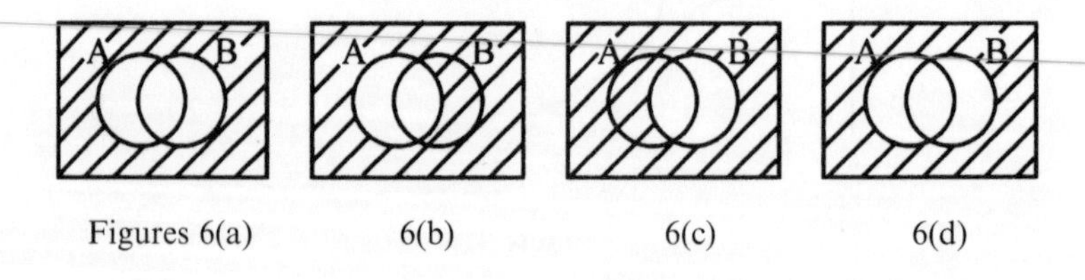

Figures 6(a) 6(b) 6(c) 6(d)

not A or B: (A ∪ B)′. **23**. It was not A: A′. **24**. It was not B: B′. **25**. (a) No; (b) No; (c) Yes. **26**. (a) Yes; (b) Yes; (c) Yes.

27.

A	B	Bulb
0	0	0
1	0	1
0	1	1
1	1	0

28.

A	Bulb
0	1
1	0

29. Yes. **30**. Yes. **31**. Yes.

Answers to Chapter 12, Some Areas and Volumes

3. 4 × 4 = 16cm^2. **4**. 6 × 16 = 96cm^2. **5**. 2 × 2 = 4cm^2. **6**. 6 × 4 = 24cm^2. **7**. 24 × 8 = 192cm^2. **8**. The volume of the 4cm cube is the same as the volume of 8 cubes with side 2cm. **9**. 64cm^3. **10**. 96 ÷ 64 = 1.5. **11**. 2^3 = 8cm^3. **12**. 24 ÷ 8 = 3. **13**. 1cm^2. **14**. 6cm^2. **15**. 1cm^3. **16**. 6 ÷ 1 = 6. **22**. 2 × *radius*. **23**. 2 × π × *radius*. **24**. 2 × *radius* × 2 × π × *radius* = 4 × π × *radius* × *radius*.

Answers to Chapter 13, Movement

1. 0.6s. **2**. 6m/s. **3**. ½ × 6 × 0.6 = 1.8m. **4**. 6 ÷ 0.6 = 10m/s/s. **5**. v^2 = 2 × 10 × 1.8 = 36; v = 6m/s. **6**. 10 × 0.1653 = 1.653m/s/s. **7**. 18 ÷ 1.653 = 11m (or 10.9m). **8**. 10 × 0.9 = 9m/s/s. **9**. 18 ÷ 9 = 2m. **10**. Pluto; g = 1m/s/s; height = 18 ÷ 1 = 18m. *Other heights*: Mercury: 4.7m; Venus: 2m; Earth: 1.8m; Mars: 4.7m; Jupiter: 6.9m; Saturn: 1.5m; Uranus: 2m; Neptune: 1.5m; Pluto: 18m. **11**. Yes! **12**. It flies off at a tangent. **13**. No! **14**. 3380N. **15**. 13520N. **16**. 135200N. **17**. 6760N. **18**. 3054 × 60 = 183240m. **19**. 183240 × 60 = 10994400m. **20**. 10994400 ÷ 1000 = 10994.4kph. **21**. 263893.78km. **22**. 24 hours.

Answers to Chapter 14, Virtual Machines

1. 6. **2**. 15. **3**. 9. **4**. 51. **5**. 11. **6**. 17. **7**. 5. **8**. 114. **9**. 11. **10**. 19. **11**. 105. **12**. 200. **13**. 6. **14**. 14. **15**. 37. **16**. 95. **17**. 4. **18**. 36. **19**. 144. **20**. 144. **21**. ±2. **22**. ±9. **23**. ±12. **24**. ±1.414. **25**. 3. **26**. 23. **27**. 35. **28**. –1. **29**. 2. **30**. 5. **31**. 6. **32**. 11.5. **33**. +5 and ÷4. **34**. 19. **35**. 28. **36**. 28. **37**. 103. **38**. ±4. **39**. ±2. **40**. ±9. **41**. ±1.732... **42**. There are many answers to this question, for example: ×6 and +3 etc. **43**. The simplest answers are inverse operations, for example: +3 and –3 or ×7 and ÷7. **45** See table 1. **46**. It prints the 3 times table. **47**. Change store B to hold the number 4. **48**. See table 2; it prints the multiplication tables up to 12 × 12. **49**. See table 3; this will print the ratio of consecutive terms of the Fibonacci Sequence, which tends to the Golden Section 1.618... **50**. Put 'Print A' just before or just after 'C = A + B'.

Table 1

A	B	C
1	3	3
2	3	6
3	3	9
4	3	12
5	3	15
6	3	18
7	3	21
8	3	24
9	3	27
10	3	30
11	3	33
12	3	36

Table 2

A	B	C
0	1	0
1	1	1
2	1	2
3	1	3
.	.	.
1	2	2
2	2	4

Table 3

A	B	C	D	E
1	1	0	0	0
1	2	2	2	1

Answers to Chapter 15, Topology

3. There is no such thing as a 2-node – or there is an infinite number of them! **4**. Both 3-nodes. **5**. Both 1-nodes. **6**. 4-nodes. **7**. Odd-numbered nodes are start or finish points of a unicursal curve. **8**. No; it has 3 places where it must start or finish. **9**. Yes. **10**. Anywhere; it starts and finishes at the same point. **11**. and **12**. No; all the nodes are odd-numbered, so there are 4 start or finish points. **13**. (0,0) (0,1) (1,1) (1,2) (2,2) (2,3) (3,3) (3,4) (4,4) (4,1) (1,3) (3,0) (0,2) (2,4) (4,0); there are other answers which work. **14**. See the first 6 pieces of question 13. **15**. See question 13. **16**. No. **17**. The solution is impossible. All the nodes are 5-nodes (ignoring the intersections of the diagonals) so the figure is not unicursal. **18**. Many solutions are possible. One is: (0,0) (0,1) (1,1) (1,2) (2,2) (2,3) (3,3) (3,4) (4,4) (4,5) (5,5) (5,6) (6,6) (6,1) (1,5) (5,3) (3,1) (1,4) (4,2) (2,6) (6,4) (4,0) (0,5)

(5,2) (2,0) (0,3) (3,6) (6,0). **21**. 1, 2, 4, 5, 8, 9. **22**. The line from 1 to 6. **23**. At least 8; 5 loops are obvious. **24**. 3. **25**. No; you do not go along paths 3, 6 or 7. **27**. Yes. **29**. No. **30**. No; you touch all the inside or all the outside, but not both. **31**. Yes.

Answers to Chapter 16, Numbers and Patterns

1. No. **2**. A square. **3**. (a) x x x Prime; (b) x x x x x x x Prime;

(c) x x x x	(d) x x x	(e) x x x x x x	(f) x x x x x x
x x x x	x x x	x x x x x	x x x x x x
Rectangle	x x x	Prime	Rectangle
	Square		

6. 3. **8**. 5. **9**. 7. **10**. 2: 1 + 3. **11**. 3: 1 + 3 + 5. **12**. 20. **13**. $10^2 = 100$. **14**. 5, 7, 11, 13, 17, 19, 23, (not 25) 29, 31, (not 35) 37. **15**. (a) 3; (b) 6. **16**. 2. **17**. 3. **18**. 6 + 4 = 10; 10 + 5 = 15. **19**. 1 + 3 = 4. **20**. 3 + 6 = 9. **21**. 6 + 10 = 16. **22**. $10^2 = 100$. **24**. 3rd diagonal. **25**. 1, 4, 9, 16. **26**. (a) 3; (b) 6; (c) 10. **27**. 5050. **28**. 5. **29**. 12 = 5 + an extra 7. **30**. 22 = 12 + an extra 10. **31**. $2 \times T_3 + T_4$; $2 \times T_4 + T_5$; $2 \times T_5 + T_6$ or, for example, the 4th pentagon number = 4th triangle number + 2 x 3rd triangle number. **32**. $2 \times 21 + 28 = 70$. **33**. $2 \times 28 + 36 = 92$. **34**. 6. **35–37**. Differences: 3, 5, 7, 9, 11. **38**. 36 + 13 = 49. **39**. Differences: 2, 3, 4, 5, 6. **40**. 21 + 7 = 28. **41**. Differences: 4, 7, 10, 13, 16. **42**. Differences: 5, 9, 13; next differences: 17, 21; next hex numbers: 28 + 17 = 45; 45 + 21 = 66. **43**. Each alternate triangle number is a hexagonal number.

Answers to Chapter 17, Triangles and Circles

Most answers are given or implied in the text. **3**. The lines meet one third of the way along each median. **16**. The triangle is always equilateral; all its sides and all its angles are equal.

Answers to Chapter 18, Base Two

1. 1, 10, 11, 100, 101, 110, 111, 1000, 1001, 1010, 1011, 1100, 1101, 1110, 1111. **2**. 10110. **3**. 2 times bigger. **4**. 3 tenths. **5**. Half. **6**. 101. **7**. 31. **8**. 1111.1. **9**. 1111. **10**. 111.1. **11**. 111. **12**. 11.1. **13**. 11. **14**. 1.1. **15**. 1.

16

Weight(oz)	8oz	4oz	2oz	1oz	Weight(oz)	8oz	4oz	2oz	1oz
1				1	9	1	0	0	1
2			1	0	10	1	0	1	0
3			1	1	11	1	0	1	1
4		1	0	0	12	1	1	0	0
5		1	0	1	13	1	1	0	1
6		1	1	0	14	1	1	1	0
7		1	1	1	15	1	1	1	1
8	1	0	0	0					

The ticker-tape message is: 'Hello how are you today'.

INDEX